THE

PRINCIPAL METEOROLOGICAL CONDITIONS
IN MICHIGAN
DURING THE YEAR 1879:

A Compilation Based upon Reports by the

METEOROLOGICAL OBSERVERS

FOR THE

STATE BOARD OF HEALTH,

AND FOR THE U. S. SIGNAL SERVICE.

COMPILED AND PREPARED IN THE OFFICE OF THE SECRETARY OF THE BOARD, LANSING, MICHIGAN.

[Reprinted from the Annual Report of the Michigan State Board of Health, for the year 1880.]

[REPRINT NO. 76.]

THE PRINCIPAL METEOROLOGICAL CONDITIONS IN MICHIGAN DURING THE YEAR 1879.

In the two Reports preceding this, were published summaries of the principal meteorological conditions in Michigan during the years 1877 and 1878. For 1879 observations were continued at the old stations, and reports were received from eight new stations, from four of them by the courtesy of the United States Signal Service.* The new stations were Marquette, Menominee, Alpena, Grand Haven, Port Huron, State House of Correction at Ionia, Washington (Macomb Co.), and the office of the State Board of Health, Lansing. The names of observers and their places of observation are given in Exhibit 7, page 306. Observations for less than half the year have not been used in this article, and observations not for the full year have not been included in the average lines made for the several localities represented in the various tables. A copy of the monthly register in use was printed on page 216 of the Report for 1878, and suggestions as to the use of meteorological data were given on pages 213–14 of that Report.

USES DESIGNED TO BE MADE OF THESE METEOROLOGICAL DATA.

The facts tabulated and otherwise embraced in this article are collected with a view to their use in studies relating to the causes of sickness and deaths in Michigan. They have been carefully digested and prepared, and are here presented in such form as to be available for use in connection with the facts respecting sickness in Michigan, grouped in the article following this, and entitled "Weekly Reports of Diseases in Michigan, in 1879." They are also available for use in connection with the article on "Diseases in Michigan in 1879," on pages 221–262 of this Report; and they will be found useful for study in connection with the records of deaths in Michigan during the same year which are returned by the Supervisors, to be published by this State under

* The following stations of the U. S. Signal Service kindly sent to this office for the year 1879 monthly reports of their regular, tri-daily observations, in many cases corrected and reduced, and for the most part on blanks supplied from this office: Alpena, Detroit, Grand Haven, Marquette, and Port Huron. The observers at these stations also made observations of ozone especially for this Board. For study in connection with reports of sickness some of the observations require a different elaboration from that given to them by the Signal Service Office. This is true especially of the record of atmospheric pressure, which for study with sickness-reports requires to be corrected for temperature and instrumental error but not to be reduced to sea level. To the Chief Signal Officer at Washington is sent each month a copy of the register of observations made at this office, and each week a copy of the weekly summary of said observations, which is published in the Lansing Republican, and a copy is also sent to every observer reporting to this office.

the title of the "Vital Statistics of Michigan." It seems probable, however, that many physicians and sanitarians may have occasion in the future to employ the evidence here placed upon record, in connection with facts which are known only to themselves, and which will only become of sufficient value to warrant their being made public by reason of the evidence herein contained.

EXHIBIT 7.—*Names of Observers whose Reports are summarized in the following Meteorological Tables and Diagrams, their Places of Observation, and the Counties and Geographical Divisions of the State, in which these Places are situated.*

NAME OF OBSERVER.	PLACE OF OBSERVATION.	COUNTY.	GEOGRAPHICAL DIVISION OF THE STATE.*
Sergt. J. T. O'Keefe, U. S. Signal Corps[a]; Sergt. J. Gilligan, U. S. Signal Corps[b]	Marquette	Marquette	Upper-Peninsular.*
Rev. A. W. Bill	Menominee	Menominee	Upper-Peninsular.
H. T. Calkins, M. D.	Petoskey	Emmet	Northern.
Sergt. Jas. A. Barwick, U. S. Signal Corps	Alpena	Alpena	North-eastern.
Sergt. S. C. Emery, U. S. Signal Corps[c]; Corpl. D. T. Waters, U. S. Signal Corps[d]	Grand Haven	Ottawa	Western.
Lee S. Cobb	Nirvana	Lake	Western.
Sergt. W. O. Bailey, U. S. Signal Corps	Port Huron	St. Clair	Bay and Eastern.
John S. Caulkins, M. D.	Thornville	Lapeer	Bay and Eastern.
Prof. R. C. Kedzie[e]	Agricultural College, near Lansing.	Ingham	Central.
J. J. Grafton, M. D., Warden	State House of Correction, Ionia.	Ionia	Central.
Frank S. Kedzie[f]; Morey S. Collier[g]; Harry B. Turner[h]	Office State Board of Health, Lansing.	Ingham	Central.
A. W. Nicholson, M. D.	Otisville	Genesee	Central.
John Bell, M. D.	Benton Harbor	Berrien	South-western.
James S. Reeves, M. D.	Niles	Berrien	South-western.
John H. Kellogg, M. D.	Battle Creek	Calhoun	Southern-central.
Lyman P. Alden, Supt.	State Public School, Coldwater	Branch	Southern-central.
A. M. Munn	Michigan Asylum for Insane, Kalamazoo.	Kalamazoo	Southern-central.
Edwin Stewart, M. D.	Mendon	St. Joseph	Southern-central.
Harrison Peters, M. D.	Tecumseh	Lenawee	Southern-central.
Prof. Lewis McLouth; Lawrence A. McLouth	State Normal School, Ypsilanti	Washtenaw	Southern-central.
Sergt. Theo. V. Van Heusen, U. S. Signal Corps[i]; Sergt. C. F. R. Wappenhaus, U. S. Signal Corps[j]	Detroit	Wayne	South-eastern.
F. W. Higgins, Supt.	Woodmere Cemetery, near Detroit.	Wayne	South-eastern.
Albert Yates, M. D.	Washington	Macomb	South-eastern.

* The counties included in each division are stated in Exhibit 1, page 227.

[a] For July, and from Sept. to Dec. [b] From Jan. to June, and for August. Sergt. Gilligan, who took charge of the station at Marquette in January, 1880, kindly sent copies of records for the first months of 1879 made by his predecessor, who had begun reporting to the State Board of Health in July, 1879. [c] From Jan. to Sept. [d] From Oct. to Dec. [e] The statements for the Agricultural College for 1879 are taken from Prof. Kedzie's "Register of Meteorological Conditions" in the Report of the State Board of Agriculture for 1879. For November and December they are copied from observations made in the office of the State Board of Health, Lansing. [f] From Jan. to May [g] From June to Aug. [h] From Sept. to Dec. [i] From Jan. to March. [j] From April to Dec.

DIAGRAMS RELATING TO METEOROLOGICAL CONDITIONS IN 1879.

The relative intensity or prevalence of the several meteorological conditions in different months is best understood when represented to the eye by curved or angular lines, the rise and fall of which accurately depict according to a fixed scale the relative intensity or prevalence of the condition. Fourteen such diagrams have been made for 1879. By means of these in connection with the diagrams for 1877 and 1878, in the summaries for those years in the Reports for 1878 and 1879, a ready comparison may be had of conditions in these years. These diagrams may also be studied in connection with the diagrams relating to sickness in a paper on "Weekly Reports of Diseases in Michigan," on later pages of this Report; and this is the main purpose of their construction. In studying these diagrams it should be remembered that the significant points are to be found in and not between the upright lines for the several months, as named at the top of these lines. Diagrams so made seem to be more easily read than those with the significant points between the upright lines. It should be remembered also that although the irregular lines crossing the diagrams from left to right do not necessarily represent the exact course or range of the several conditions from month to month (except in the cases in which there may have been a *uniform* increase or decrease from one month to another), yet they serve to guide the eye from the significant point in the upright line for one month to that in the line for the next month.

As seems most proper, it has been sought, by making the heads and notes to these diagrams as nearly self-explanatory as possible, to save the reader needless trouble in turning from page to page in order to find to what the diagram relates or how it is to be read. The following explanations, however, may be repeated, slightly modified, from the Report of 1879, to aid those not familiar with this method of graphic statement:—

The intersection of the irregular line crossing the diagram from left to right and representing conditions at any locality, with the perpendicular line for any month marks the degree or the amount intended to be indicated for that month at that locality. This degree or amount may readily be determined by following the nearest horizontal line to the number stated in the left-hand margin of the diagram, and, if necessary, adding or subtracting a quantity proportional to the distance of the said point of intersection above or below that horizontal line. In the tables these degrees or amounts are definitely stated for each condition at each locality, and to the tables reference may easily be made from any diagram by considering the subject of the diagram. For convenience of reference, the diagrams are in most cases printed opposite the tables from which they are made, and which they are designed to illustrate. Diagrams XI., XII., and XIII., relating to the direction of the wind, are somewhat differently constructed and require separate explanation.

DIAGRAMS XI., XII., AND XIII., RELATIVE TO THE DIRECTION OF THE WIND.

The figures or separate groups of lines in these diagrams are designed to indicate the number and the proportion of regular observations, at 7 A. M., 2 P. M., and 9 P. M. daily, at which the wind was blowing from each of the eight principal points of compass at the places and for the periods of time stated in the margin. The points of compass are represented as is usually done on maps, namely,—north, by the top of the page; south, by the bottom; east by the right-hand; and west, by the left-hand side. Each figure consists of lines drawn to a common center from some or all of the following directions on the page, and indicating that at the times of observation the wind blew from points of the compass as follows: Lines toward the common center from the top of the page indicate that the wind blew from the north; from the right-hand side, that the wind was from the east; from the bottom of the page, that it was from the south; from the left-hand side, that it was from the west; from the upper left-hand corner, that it was from the north-west; from the upper right-hand corner, that it was from the north-east; from the lower right-hand corner, that it was from the south-east; from the lower left-hand corner, that it was from the south-west. The length of each line denotes the number of regular observations at which the wind blew from the given direction, .01 of an inch being the unit, or the length of line for one observation. The circles indicate calms, the number of regular observations at which there was no wind being denoted by the length of the radius of the circle drawn about the point of convergence of the lines for a given place or period of time, the length for one observation being, as before, .01 of an inch.

ERRORS IN ENGRAVING DIAGRAMS.

Great care has been taken to have the diagrams carefully drawn to scale. In the engraving of some of the plates, however, slight errors have occurred.

Diagrams I.-XIV., relating to meteorological conditions, were drawn to a scale twice as great as stated at the foot of the diagrams, with instructions to the photo-engraver to reduce them one-half in length and one-half in width. Diagram XIII., page —, was, however, so reduced as to be twenty-two-fiftieths of an inch less in length than designed; and the other dimensions are therefore proportionally less. This error makes the longer line for each direction of the wind from one to three hundredths of an inch shorter than designed in the statement of the scale. The error in Diagram XI., page ..., is about two-thirds as great; and in Diagram XII., page ..., about one-fourth as great. The exact number of observations for each direction of the wind is stated in Tables XI., XII., and XIII., pages ..., ..., ... Diagrams IX. and X., also, pages ... and ..., relating to the velocity of the wind at Lansing, are about twenty-one-fiftieths of an inch shorter than designed, and in them also the significant dimensions are proportionally less than stated in the "scales." The other diagrams vary not to exceed one-fiftieth to one-tenth of an inch in length from what they were designed. In any close comparisons of conditions, reference should, of course, be made to numbers in the tables.

THE MANNER OF USING THESE DIAGRAMS.

Persons unaccustomed to the use of diagrams, or not familiar with this particular study, sometimes remark that they are unable to follow all the lines, and therefore think they fail to get the ideas intended to be conveyed, which is probably the fact, for the reason that they look in a wrong direction. To such persons it is proper to explain that while it is desirable in some diagrams to endeavor to follow each line through its entire course, some of these diagrams are planned to show by other methods other facts believed to be of fully as much importance; as, for instance, in the two diagrams relating to the average temperature and the absolute humidity, it is shown that at several stations quite widely separated, the temperature and the absolute humidity did not vary much from the averages at the other stations. That many lines disappear behind others is an important part of the evidence, because showing that the conditions were the same at many places, and even at places widely remote from each other. If only two or three stations at which the conditions were sufficiently different to permit of following their lines distinctly, apart from the others, had been represented on these diagrams, nothing would have been gained for our present main purpose (which is to get a good knowledge of the meteorology of the State as a whole, for use in study in connection with the reports of sickness which thus far are represented by diagrams and studied for the State as a whole); but much would have been lost because those diagrams gain very great strength from the general concurrence of the evidence (much as a rope is strengthened by the entwisting of its separate twines), which is such a strongly marked feature in each of them.

It may be stated as generally applicable to the diagrams in this part of the Report, that one important idea intended to be conveyed first of all, is the general tendency of the evidence. If special details are wanted, the reader should usually look for them in the tables which accompany the diagrams, although in some cases it is possible to trace details in the diagrams. In most of the diagrams except such as the one on average temperature and the one on absolute humidity, in which the average of many stations is sufficiently manifest without a special line to show it, the average for all stations or for many stations, is represented by a line of xxxx's, which can be followed with the eye without difficulty. It is this average line or the average evidence of a diagram of a meteorological condition which should be studied in connection with each particular line which in the diagrams further on in this Report represents the average rise and fall of a disease in the State, considered as a whole.

In the diagrams representing distinct diseases it is usually designed to represent only so many on one diagram as will permit of the line for each disease being readily followed. By this means clear ideas in one direction are gained, though something is lost in the way of showing the natural relationships of the several diseases, but where it is practicable this is also kept in mind, and when the necessity is here pointed out, the reader can study that subject by picking out from each diagram the diseases which are evidently related to those represented on other diagrams. Similar remark is applicable to the diagrams of meteorological conditions, many of which it may be seen bear close relations to each other.

EXHIBIT 8.—*Statements of Meteorological Conditions in the Year and in each Month of the Year 1879, Compared with Annual and Monthly Averages for 1878 and for several Stated Periods of Years,—from Observations by Prof. R. C. Kedzie, at the State Agricultural College*, near Lansing, Michigan.*

METEOROLOGICAL CONDITIONS.	1879 COMPARED WITH AVERAGES FOR PREVIOUS YEARS.		In 1879	METEOROLOGICAL CONDITIONS.	1879 COMPARED WITH AVERAGES FOR PREVIOUS YEARS.		In 1879
	No. of Years Averaged, ending with 1878.	More (+), or Less (-), in 1879 than the Average for Previous Years.	More (+), or Less (-), than in 1878.		No. of Years Averaged, ending with 1878.	More (+), or Less (-), in 1879 than the Average for Previous Years.	More (+), or Less (-), than in 1878.
YEAR 1879.*				YEAR 1879.—*Continued.*			
Av. Temp'ature	15	+0.24° F.	-1.41° F.				
Range of Temp†	6	+1°	+10°				
Av. Monthly Range of Temp.†	6	+3°	+10°	Rainfall	15	-3.62 in.	-5.17 in.
Av. Daily Range of Temp.‡	9	-0.65°	+1.34°	Day Ozone §	7	-.52°	-.26°
Cloudiness	15	-5 per ct.	-4 per ct.	Night Ozone §	7	-.75°	-.06°
				Atmospheric Pressure	4	+.067 in.	+.042 in.
JANUARY.				FEBRUARY.			
Av. Temp.	15	-3.21°	-9.92°	Av. Temp.	15	-3.97° F.	-7.67° F.
Range of Temp.†	6	+2°	+10°	Range of Temp†	6	-12°	-15°
Av. Daily Range of Temp.‡	9	+2.18°	+4.65°	Av. Daily Range of Temp.‡	9	-4.37°	-3.09°
Cloudiness	15	-6 per ct.	-8 per ct.	Cloudiness	15	+10 per ct.	+17 per ct.
Rainfall	15	-1.27 in.	-.63 in.	Rainfall	15	-27 in.	-1.31 in.
Day Ozone §	7	-.87°	+.26°	Day Ozone §	7	-.43°	-.46°
Night Ozone §	7	-1.30°	0	Night Ozone §	7	-.90°	-.32°
Atmospheric Pressure	4	+.071 in.	+.031 in.	Atmospheric Pressure	4	+.085 in.	+.081 in.
MARCH.				APRIL.			
Av. Temp.	15	+1.70°	-7.71°	Av. Temp.	15	-1.14°	-5.71°
Range of Temp.†	6	-4°	+8°	Range of Temp†	6	+7°	+23°
Av. Daily Range of Temp.‡	9	-2.19°	+0.68°	Av. Daily Range of Temp.‡	9	+2.59°	+4.96°
Cloudiness	15	-9 per ct.	-16 per ct.	Cloudiness	15	-14 per ct.	-15 per ct.
Rainfall	15	-1.26 in.	-1.55 in.	Rainfall	15	-1.24 in.	-2.51 in.
Day Ozone §	7	-.90°	-.75°	Day Ozone §	7	-.94°	-1.37°
Night Ozone §	7	-1.66°	-.64°	Night Ozone §	7	-.92°	-.53°
Atmospheric Pressure	4	+.157 in.	+.075 in.	Atmospheric Pressure	4	+.076 in.	+.175 in.

* For November and December, 1879, the observations were made by Harry B. Turner, at the office of the State Board of Health, Lansing.

† By registering thermometers, set at 7 A. M., and recorded at 7 A. M. for the preceding calendar day.

‡ By observations at 7 A. M., 2 P. M., and 9 P. M., daily.

§ Degrees, by scale of 10 degrees of coloration of Schönbein's test-paper, exposed from 7 A. M. to 2 P. M., for the day observation; and from 9 P. M. to 7 A. M., for the night observation.

EXHIBIT 8.—CONTINUED.—*Meteorological Conditions in Months of the Year 1879, Compared with Averages for Corresponding Months in Preceding Years.*

METEOROLOGICAL CONDITIONS.	1879 COMPARED WITH AVERAGES FOR PREVIOUS YEARS.		In 1879	METEOROLOGICAL CONDITIONS.	1879 COMPARED WITH AVERAGES FOR PREVIOUS YEARS.		In 1879
	No. of Years Averaged, ending with 1878.	More (+), or Less (-), in 1879 than the Average for Previous Years.	More (+), or Less (-), than in 1878.		No. of Years Averaged, ending with 1878.	More (+), or Less (-), in 1879 than the Average for Previous Years.	More (+), or Less (-), than in 1878.
MAY.				JUNE.			
Av. Temp.	15	+0.66° F.	+4.19° F.	Av. Temp.	15	-2.40° F.	+1.94° F.
Range of Temp.†	6	+5°	+18°	Range of Temp†	6	+8°	+7°
Av. Daily Range of Temp.‡	9	+2.27°	+5.60°	Av. Daily Range of Temp.‡	9	-0.56°	+0.80
Cloudiness	15	-12 per ct.	-14 per ct.	Cloudiness	15	-7 per ct.	-2 per ct.
Rainfall	15	-.40 in.	-.99 in.	Rainfall	15	-1.00 in.	-.28 in.
Day Ozone§	7	-.93°	-1.18°	Day Ozone§	7	-.55°	-.74°
Night Ozone§	7	-1.40°	-1.42°	Night Ozone§	7	-.84°	-.97°
Atmospheric Pressure	4	+.081 in.	+.103 in.	Atmospheric Pressure	4	+.075	+.043
JULY.				AUGUST.			
Av. Temp.	15	+2.02°	+0.99°	Av. Temp.	15	+0.87°	-0.15°
Range of Temp.†	6	0	-1°	Range of Temp.†	6	+7°	+11°
Av. Daily Range of Temp.‡	9	-0.16	+2.52	Av. Daily Range of Temp.‡	9	+1.19°	+1.35°
Cloudiness	15	-11 per ct.	-1 per ct.	Cloudiness	15	-13 per ct.	-5 per ct.
Rainfall	15	-1.14 in.	-.77 in.	Rainfall	15	-1.03	-.24
Day Ozone§	7	+.51°	+.80°	Day Ozone§	7	-.07°	+.26°
Night Ozone§	7	-.03°	+.39°	Night Ozone§	7	+.01°	+.24°
Atmospheric Pressure	4	+.007 in.	-.028 in.	Atmospheric Pressure	4	+.068 in.	+.090 in.
SEPTEMBER.				OCTOBER.			
Av. Temp.	15	-4.13°	-6.94°	Av. Temp.	15	+9.83°	+8.95°
Range of Temp.†	6	0	-3°	Range of Temp.†	6	+12°	+11°
Av. Daily Range of Temp.‡	9	-3.33°	-1.30°	Av. Daily Range of Temp.‡	9	-1.54°	-0.54°
Cloudiness	15	-3 per ct.	+2 per ct.	Cloudiness	15	-4 per ct.	+1 per ct.
Rainfall	15	+.20 in.	-.24 in.	Rainfall	15	-.74	-.42
Day Ozone§	7	+.20°	+.64°	Day Ozone§	7	-.17°	+.61°
Night Ozone§	7	+.26°	+.93°	Night Ozone§	7	-.86°	+.84°
Atmospheric Pressure	4	+.076 in.	-.005 in.	Atmospheric Pressure	4	+.134 in.	+.080 in.
NOVEMBER.*				DECEMBER.*			
Av. Temp.	15	+3.01°	+1.93°	Av. Temp.	15	+2.46°	+6.17°
Range of Temp.†	6	+9°	+25°	Range of Temp.†	6	+7°	+23°
Av. Daily Range of Temp.‡	9	-2.16°	-0.67°	Av. Daily Range of Temp.‡	9	-1.83	+0.81
Cloudiness	15	0	0	Cloudiness	15	+8 per ct.	-5 per ct.
Rainfall	15	+2.67 in.	+2.39 in.	Rainfall	15	+1.73	+1.29
Day Ozone§	7	-.46°	+.36°	Day Ozone§	7	-1.61°	-1.58°
Night Ozone§	7	-.08°	+1.49°	Night Ozone§	7	-1.35°	-.87°
Atmospheric Pressure	4	-.006 in.	-.040 in.	Atmospheric Pressure	4	-.025 in.	-.101 in.

* For November and December 1879 the observations were made by Harry B. Turner, at the office of the State Board of Health, Lansing.

† By registering thermometers. ‡ By observations at 7 A. M., 2 P. M. and 9 P. M., daily.

§ Degrees, by scale of 10 degrees of coloration of Schönbein's test-paper, exposed from 7 A. M. to 2 P. M., for the day of observation; and from 9 P. M. to 7 A. M., for the night observation.

METEOROLOGICAL CHARACTERISTICS OF THE YEAR 1879.

By the comparison in Exhibit 8, pages 309–310, of conditions at the Agricultural College in 1879, with averages for preceding periods of years, it would seem that the average temperature, the ozone, and the atmospheric pressure varied but little from those averages for those periods. The cloudiness was about 5 per cent less, and the rainfall 3.62 inches less than the average for the preceding 15 years. In November and December the rainfall was greater than the average for the preceding 15 Novembers and Decembers.

TEMPERATURE.

The average temperature seems to have been lower in 1879 than in 1878 from January to April, inclusive, and in September; and higher in May, June, October, November, and December. Compared with the average for 15 preceding years the average temperature at the Agricultural College in 1879 was considerably lower in January, February, and September, and considerably higher in July, October, November and December. The average temperature at the Agricultural College in October, 1879, was more than 3° higher than that of any October in the preceding 15 years, and more than 8° higher than the average October temperature for that period.

The unusual high temperature in the first part of October, 1879, is well shown by the following record of temperature of water in Cedar River, near the Agricultural College, taken from the 6th to the 9th of October in each of the years 1874–1880, and kindly contributed by Prof. R. C. Kedzie.

EXHIBIT 9.—*Temperature of Water in Cedar River, taken October 6–9, for the Years 1874–80.*

	Average, 7 Years, 1874-80.	1874.	1875.	1876.	1877.	1878.	1879.	1880.
Temp. of water in Cedar River, taken from 6th to 9th of Oct., Degrees Fahr.	54°.6	54°	50°	46°	50°	60°	68°	54°

The following record of the mean October temperature for each of the 10 years, 1870–9, at Woodmere Cemetery, near Detroit, is by Superintendent F. W. Higgins:

EXHIBIT 10.—*Average October Temperature.*

	Aver'ge, 10 Yrs., 1870–9.	1870.	1871.	1872.	1873.	1874.	1875.	1876.	1877.	1878.	1879.	1880.
Mean October temperature	49.81	53.9	52.0	49.3	46.7	49.5	44.8	44.3	51.8	49.4	56.4	......

The average temperature for the year 1879, at the Agricultural College, varied less than one degree from the average for the preceding 15 years. The average temperature for 19 stations in 1879 was the same for both September and October, viz., 57°.43.

The average daily range of the 7 A. M., 2 P. M., and 9 P. M. observations at the Agricultural College varied less than one degree for the year 1879 from the average for the preceding 9 years; for February, March, September, and November it was more than 2° less than the average for the corresponding months of the preceding 9 years; and for January, April, and May it was more than 2° greater than the average for the corresponding months in the same period. The average temperature in 1879, as observed at the office of the State Board of Health, Lansing, was about 2°.20 higher than as observed at the State Agricultural College. The average for the 10 months, January to October inclusive, during which observations were made at the College, was 2°.41 higher at the office of the State Board of Health than at the College. The difference is greatest in the cold months. This difference in the temperature of two places so near may be in part explained by the fact that the instruments at the office of the Board of Health are placed too near the capitol building, are influenced by radiation of heat from the building, and by warm currents of air ascending near the building; and also in part by the fact that at the office of the State Board of Health the observations, broken up by moving into the new capitol, were not commenced for 1879 till January 16, after the coldest weather of the month and of the year was over. It is probable that the average temperature given for November and December, 1879, at the College, in Exhibit 12, being taken from the observations at the State Board of Health office, are somewhat too high; and that the average for the year 1879 is thereby made a little too high. By Table III., pages 322-3, giving the extremes of temperature for the year, and by months of the year 1879, it seems that for the year the range of temperature at the College and at the office of the State Board of Health was the same; but it will be noticed that in most months the minimum at the College was lower than at the office of the State Board of Health. The 20° given as the minimum for January at the office of the State Board of Health, was the only observation made during the first six days of that month. The maximum is one or two degrees higher at the College in several months, and is from one to six degrees higher at the office of the State Board of Health in other months.

The following comments relative to temperature are taken from the meteorological registers for months named:—

DEPTH OF GROUND FROZEN.

Place.	Date.	Depth.
Thornville.	January 31.	Open places about 10 inches; sheltered 0.
"	February 28.	Exposed knolls, 18 inches; sod, 8 or 10 inches; in the woods scarcely at all.
Tecumseh.	(March register.)	About 15 inches the past winter.
Thornville.	March 31.	Mostly out of ground; ice still on lakes and ponds; snowbanks on declivities.
Thornville.	(November register.)	But little frost in ground.
Thornville.	(December register.)	Not deeply.

The ground is entirely bare of snow (Feb. 1), and but a few inches have covered it at any time. This may account for the depth of ice on the bay.—*Rev. A. W. Bill, of Menominee, on January (1879) register*.

Navigation opened March 20.—*S. C. Emery, Sergt. Signal Corps, U. S. A.. Grand Haven.*

Ice out of river April 11.—*Jas. A. Barwick, Sergt. Signal Corps, U. S. A., Alpena.*

Trees covered with rime at Thornville, Jan. 12, Feb. 19 (in the morning), and Dec. 17 (loaded with ice), 1879.

EXHIBIT 11.—*Latitude and Longitude, Elevation above Sea Level, and the Average Temperature, and Average Barometric Pressure in 1879, at 23 Meteorological Stations in Michigan,—the names of the Stations being arranged in order by Latitude, highest first.*

LOCALITIES IN ORDER OF LATITUDE,—THOSE FARTHEST NORTH, FIRST.	Latitude, North.	Longitude West from Greenwich.	Altitude (Approximate), above Sea Level.—Feet.	Height of Mercury in Cistern of Barometer, above Sea Level.	Average Temperature, 1879.—Degrees F.	Average Atmospheric Pressure, 1879.—Inches of Mercury, Corrected for Temp.
Marquette	46°33′	87°36′	636.22	666.33	43.14	29.267
Petoskey	*45°22′	*84°57′	§637.	---------	42.20	---------
Menominee	*45°6′	*87°35′	---------	---------	---------	---------
Alpena	45°5′	83°28′	587.9	609.5	42.39	29.340
Nirvana	*43°54′	*85°42′	‖980	---------	44.85	29.054
Otisville	43°13′	83°31′	820	---------	46.58	---------
Grand Haven	43°5′	86°18′	593.3	616.3	48.08	29.359
Ionia	†42°59′	85°4′	688.10	---------	---------	---------
Port Huron	42°58′	82°29′	600	630.	45.97	29.341
Thornville	*42°55′	*83°12′	‖975	980.	48.11	---------
Agricultural College, near Lansing	42°44′	84°29′	834.	---------	---------	---------
Lansing	‡42°44′	†84°33′	§800	---------	49.14	29.044
Washington	42°40′	83°	746.33	752.33	46.44	29.135
Detroit	42°20′	83°2′	593	645.	48.42	29.348
Battle Creek	*42°20′	*85°11′	§800	---------	50.17	28.655
Woodmere Cemetery, near Detroit	42°18′	83°5′	588	590.	46.68	29.387
Kalamazoo	*42°18′	*85°35′	§782	---------	47.57	29.099
Ypsilanti	*42°15′	*83°36′	§700	---------	49.39	28.942
Benton Harbor	*42°8′	*86°28′	§582	---------	---------	---------
Mendon	*42°2′	*85°29′	§872	---------	47.08	---------
Tecumseh	*42°1′	*83°57′	§821	---------	46.59	29.041
Coldwater	*41°58′	*85°0′	§989	---------	48.35	---------
Niles	*41°51′	*86°16′	§695	---------	48.45	---------

* Estimated from lines on a map of Michigan issued by the General Land Office, Department of the Interior, 1878. For stations not thus marked the latitude and longitude were stated by the observer on the meteorological registers received.

† The exact latitude and longitude of the astronomical post at Ionia is 42°58′ 52″.53 N., and 85°03′ 49″.20 W.

‡ The exact latitude and longitude of the astronomical post placed in the ground near the new Capitol at Lansing, by the U. S. Lake Survey in 187., as determined by observations then made, is 42°43′ 53″.11 N., and 84°33′19″.68 W.

§ Estimated from data on "Railroad Profiles," pages 179-187, Annual Report of the State Board of Health for 1878.

‖ Estimated from data on Tackabury's Atlas of the State of Michigan.

EXHIBIT 12.—*Comparison of the Average Temperature during the Year and during each Month of the Year 1879, with the Annual and Monthly Averages for the Year 1878, and with the Average for the fifteen Years 1864–78.—Observations made by Prof. R. C. Kedzie, at the State Agricultural College*, near Lansing, Michigan.*

YEARS, ETC.	AVERAGE (MEAN) TEMPERATURE,—DEGREES FAHR.												
	Annual Av.	Jan.	Feb.	Mar.	Apr.	May.	June.	July.	Aug.	Sept.	Oct.	Nov.*	Dec.*
Av. for 15 years, 1864–78	46.64	22.40	24.37	31.49	45.98	58.10	68.42	72.01	69.13	60.34	47.45	35.21	25.00
1878	48.29	29.11	28.07	40.90	50.55	54.57	64.08	73.04	70.15	63.15	48.33	36.29	21.29
1879*	46.88	19.19	20.40	33.19	44.84	58.76	66.02	74.03	70.00	56.21	57.28	38.22	27.46
In 1879 **Higher** than Av. 15 years, 1864–78	.24	......	......	1.70	......	.66	......	2.02	.87	......	9.83	3.01	2.46
In 1879 **Lower** than Av. 15 years, 1864–78	......	3.21	3.97	......	1.14	......	2.40	......	......	4.13	......	......	......
In 1879 **Higher** than in 1878	......	......	......	......	......	4.19	1.94	.99	......	......	8.95	1.93	6.17
In 1879 **Lower** than in 1878	1.41	9.92	7.67	7.71	5.71	......	......	......	.15	6.94	......	......	......

* For November and December, 1879, the observations were made by Harry B. Turner, at the office of the State Board of Health, Lansing.

EXHIBIT 13.—*Comparison, by Year and Months, of the Average Daily Temperature at 19 Localities (in Michigan) in 1879 with the Average in 1878 at 14 Localities.**

YEARS, ETC.	AVERAGE TEMPERATURE,—DEGREES FAHR.												
	Annual Av.	Jan.	Feb.	Mar.	April.	May.	June.	July.	Aug.	Sept.	Oct.	Nov.	Dec.
Av. at 14 Stations in 1878	49.24	27.17	29.75	41.46	52.27	54.73	65.18	74.22	70.92	63.99	50.13	38.34	22.74
Av. at 19 Stations in 1879	46.82	20.86	20.69	33.08	44.29	58.03	64.70	73.16	68.99	57.43	57.43	36.80	26.41
In 1879 **Greater** than in 1878	......	......	......	......	......	3.30	......	......	......	......	7.30	......	3.67
In 1879 **Less** than in 1878	2.42	6.31	9.06	8.38	7.98	......	.48	1.06	1.93	6.56	......	1.54	......

* Thirteen of the stations, Petoskey, Nirvana, Otisville, Thornville, Niles, Battle Creek, Coldwater, Kalamazoo, Mendon, Tecumseh, Ypsilanti, Detroit, and Woodmere Cemetery, were the same for both years; Benton Harbor was included in the average for 1878 but not in that for 1879; and five, Marquette, Alpena, Grand Haven, Port Huron, Lansing, and Washington (Macomb Co.), were included in the average for 1879 but not in that for 1878.

DIAGRAM NO. I.—TEMPERATURE, BY MONTHS IN 1879.

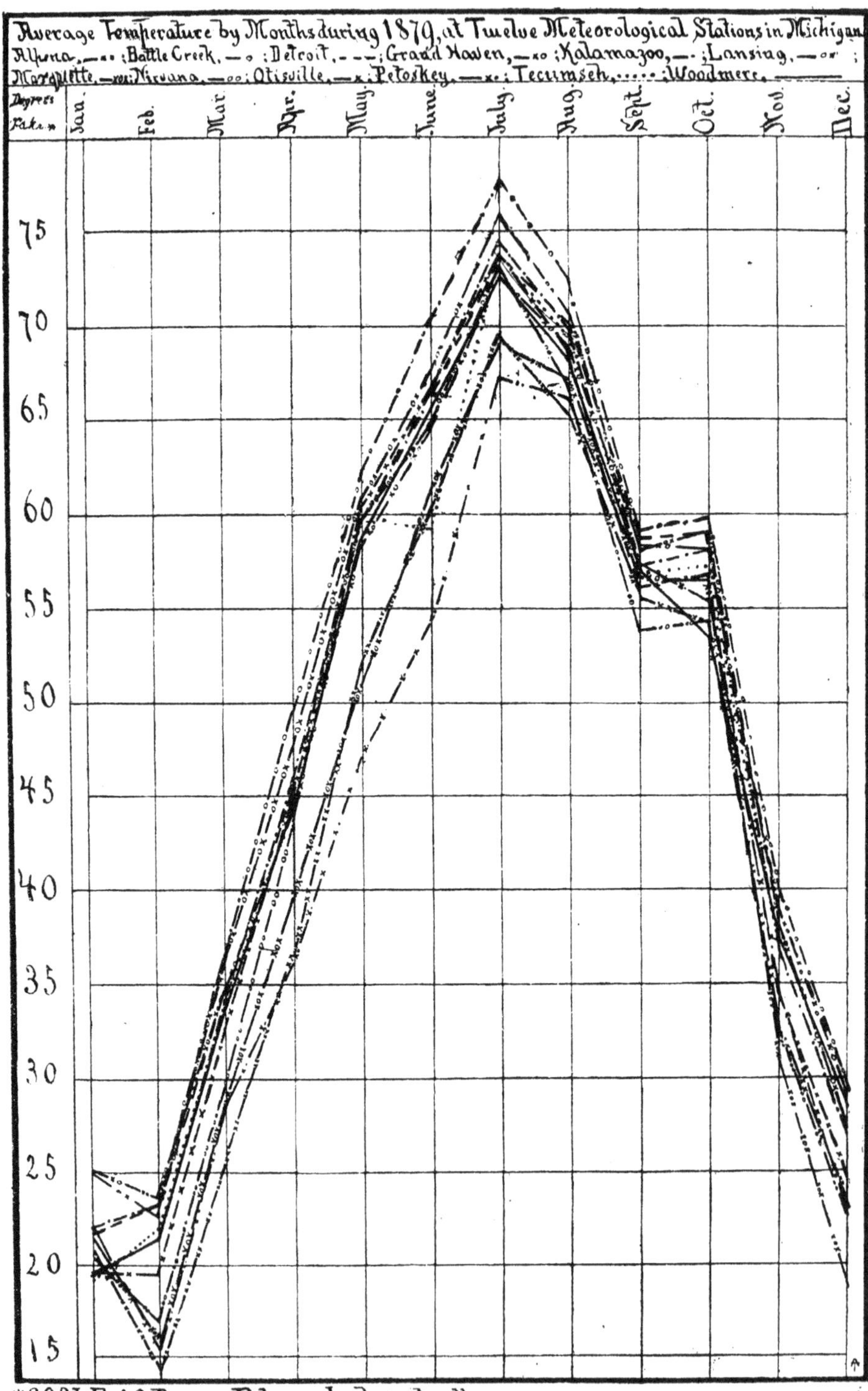

TABLE I.—*Average Temperature, in Degrees Fahr., for the Year, and for each Month of the Year 1879, at each of 23 Stations in Michigan, and also the Average for 19 of the Same Stations—as indicated by the Average of Observations made Daily at 7 A. M., 2 P. M., and 9 P. M., by Observers* for the State Board of Health, and for the U. S. Signal Service.*

STATIONS IN MICHIGAN.*	Geographical Divisions of the State.†	TEMPERATURE, IN DEGREES FAHR.													
		YEAR.		MONTHS, 1879.											
		1878.	1879.	Jan.	Feb.	Mar.	Apr.	May.	J'ne.	J'ly.	Aug.	Sept.	Oct.	Nov.	Dec.
AV. FOR 19 STATIONS ‡	----------	a	46.82	20.86	20.69	33.08	44.29	58.03	64.70	73.16	68.99	57.43	57.43	36.80	26.41
Marquette*	U. P.†	-----	43.14	21.38	15.83	28.63	39.67	50.81	61.53	68.90	67.22	56.58	56.49	32.13	18.55
Menominee	U. P.	-----	b	16.43	13.07	27.62	38.92	56.16	65.72	74.10	69.72	58.40	57.14	33.57 [f]	-----
Petoskey	N.	46.80	42.20	22.05	15.31	28.57	36.09	47.00	54.21	67.32	66.14	57.30	55.37	34.52	22.53
Alpena	N. E.	-----	42.39	20.66	14.12	25.37	36.74	52.22	60.17	69.57	65.25	55.54	54.22	32.79	22.08
Grand Haven	W.	-----	48.08	25.10	23.57	34.73	44.32	58.34	64.56	72.70	68.88	58.34	58.06	39.40	29.09
Nirvana	W.	47.15	44.85	20.52	16.77	29.30	43.46	59.48	64.86	73.18	66.92	53.77	54.39	32.76	22.75
Port Huron	B. & E.	-----	45.97	20.41	19.82	31.89	41.39	55.74	62.73	71.66	68.26	57.60	57.49	37.02	27.61
Thornville	B. & E.	49.09	48.11	20.46	20.64	33.66	45.40	60.60	67.29	75.54	71.04	58.33	59.10	37.30	27.94
Agr'l College	C.	48.29	c	19.19	20.40	33.19	44.84	58.76	66.02	74.03	70.00	56.21	57.28	-----	------
Ionia	C.	-----	d	-----	-----	-----	-----	60.51	67.21	75.81	70.34	57.61	55.82	34.61	23.99
Lansing	C.	-----	49.14	25.03 [o]	22.49	36.27	47.54 [h]	60.88 [h]	67.71	75.86	70.65	58.11	59.50	38.22	27.46
Otisville	C.	48.22	46.58	19.53	19.33	32.76	44.92	60.52	66.79	73.85 [i]	68.57 [i]	56.14 [j]	56.69 [i]	34.97	24.84 [k]
Benton Harbor	S. W.	51.03	e	22.23 [k]	24.80 [g]	34.99 [h]	46.87 [g]	-----	-----	74.87 [j]	72.51 [h]	62.69 [l]	-----	33.73 [m]	29.09 [i]
Niles	S. W.	49.70	48.45	19.37	22.74	34.33	48.68	60.73	66.71	74.44	68.95	57.82	59.19	39.60	28.83
Battle Creek	S. C.	51.46	50.17	22.05	23.33	36.37	49.36	62.32	70.43	77.66	72.40	58.98	59.80	40.04	29.25 [k]
Coldwater	S. C.	49.97	48.35	21.62 [i]	24.40	36.23	47.03	56.39 [g]	68.71 [k]	74.16	70.95	57.76	58.56	37.27	27.16
Kalamazoo	S. C.	49.19	47.57	19.47	21.38	34.46	45.86	59.68	66.68 [k]	74.58	69.78	57.32	57.99	36.84	26.85
Mendon	S. C.	48.60	47.08	17.64 [n]	21.93 [n]	34.37 [n]	46.14 [h]	59.47 [k]	64.36 [k]	74.10 [i]	68.36 [k]	57.20 [j]	57.12 [n]	37.11 [j]	27.17
Tecumseh	S. C.	49.43	46.59	18.97	22.02	33.94	44.68	59.34	59.18	73.70	69.39	56.81	57.22	36.83	27.04
Ypsilanti	S. C.	50.25	49.39	19.87 [p]	22.98	35.74	47.00	62.92	67.26 [j]	74.16 [i]	71.24 [h]	61.55 [k]	61.14 [k]	40.22 [n]	28.54 [g]
Detroit	S. E.	49.87	48.42	21.68	23.17	34.53	45.38	59.72	66.30	73.62	70.13	58.73	59.13	39.20	29.45
Woodmere Cemetery.	S. E.	48.61	46.68	19.39	21.54	33.75	44.51	58.58	65.56	72.60	68.05	56.92	53.36	37.37	28.48
Washington	S. E.	-----	46.44	21.18	21.67 [n]	33.54	43.37	57.78	64.21	72.47	68.54	56.32	56.34	35.52	26.35

* The names of observers, their places of observation, and the counties in which these places are situated, are stated in Exhibit 7, page 306.

† The names of divisions, and the counties in each, are stated in Exhibit 1, page 227.

‡ This line is an average for only the 19 stations from which statements were received for every month of the year.

In the columns from January to December, inclusive, the letters f, g, h, etc., stand directly under the numbers from which they refer to the notes below.

a The average for 15 localities in 1878 is 49.18°. b For 11 months, 46.44°. c Including the observations at Lansing for Nov. and Dec., the average of this line is 46.88°; the average for the 10 months, Jan. to Oct. inclusive, is 49.99°. d For 8 months, 55.74°. e For 9 months, 44.64°. f For 24 days. g For 25 days. h For 26 days. i For 30 days. j For 28 days. k For 29 days. l For 16 days. m For last 15 days. n For 27 days. o For the last 25 days, and the first days of the month were the coldest. p For 20 days.

The lines for 12 representative stations in Table I. are graphically represented in Diagram I., on the opposite page. Comments on the diagrams are printed on pages 307–9.

AVERAGE DAILY RANGE OF TEMPERATURE.

Most beginners in sanitary science assume that sudden changes in the temperature are very productive of sickness. After a study of the subject from time to time during more than ten years, the writer has found that the positive evidence of the truth of such an assumption is not easy to get. There is no difficulty in finding evidence that long-continued cold and long-continued heat are followed by increased sickness; this is apparent on comparing Diagram I. in this article with Diagram 1 in the article following—relative to sickness as reported weekly to the State Board of Health,—the group of diarrheal diseases are apparently closely related to long-continued high temperature; the diseases of the air-passages seem to bear a close relation to a low temperature; there are, however, few if any diseases which follow very closely the curve representing the average daily range of temperature; it is worthy of note however that one disease, intermittent fever, which has been thought (by the writer at least) to have a relation to range of temperature, seems to be more than usually prevalent in months when the average daily range of temperature is greater than the average for the year, and to be less than usually prevalent in months when the average daily range of temperature is less than the average for the year. The subject may be most conveniently studied by noticing the line of xxx's, being the average for six stations, in Diagram II., representing the average daily range at seven stations in Michigan, and comparing this line with the uppermost line in Diagram 3, in the article following this, entitled Weekly Reports of Diseases.

It may be well to bear in mind in relation to this subject that intermittent fever is a disease of the warm season in this State, that it becomes more and more prevalent as we go South until in the extreme Southern States it is extremely common and severe, that the daily range of temperature is greatest in the warm months in this State, that it is greater in the warmer climates of the Southern States, that this is especially true on the low lands, and that it is exposure during the day and during the night on the low lands in the South which is believed to be so uniformly productive of chills and fever.

If periodic chills and fever is produced or greatly favored by excessive periodic changes in the temperature of the atmosphere, then the reason is supplied why exposure on the hot low lands during the day and return to the warm high lands at night is not so productive of ague as is exposure to the cold night-air of the low lands.

DIAGRAM No. II.—RANGE OF TEMPERATURE, MONTHS IN 1879.

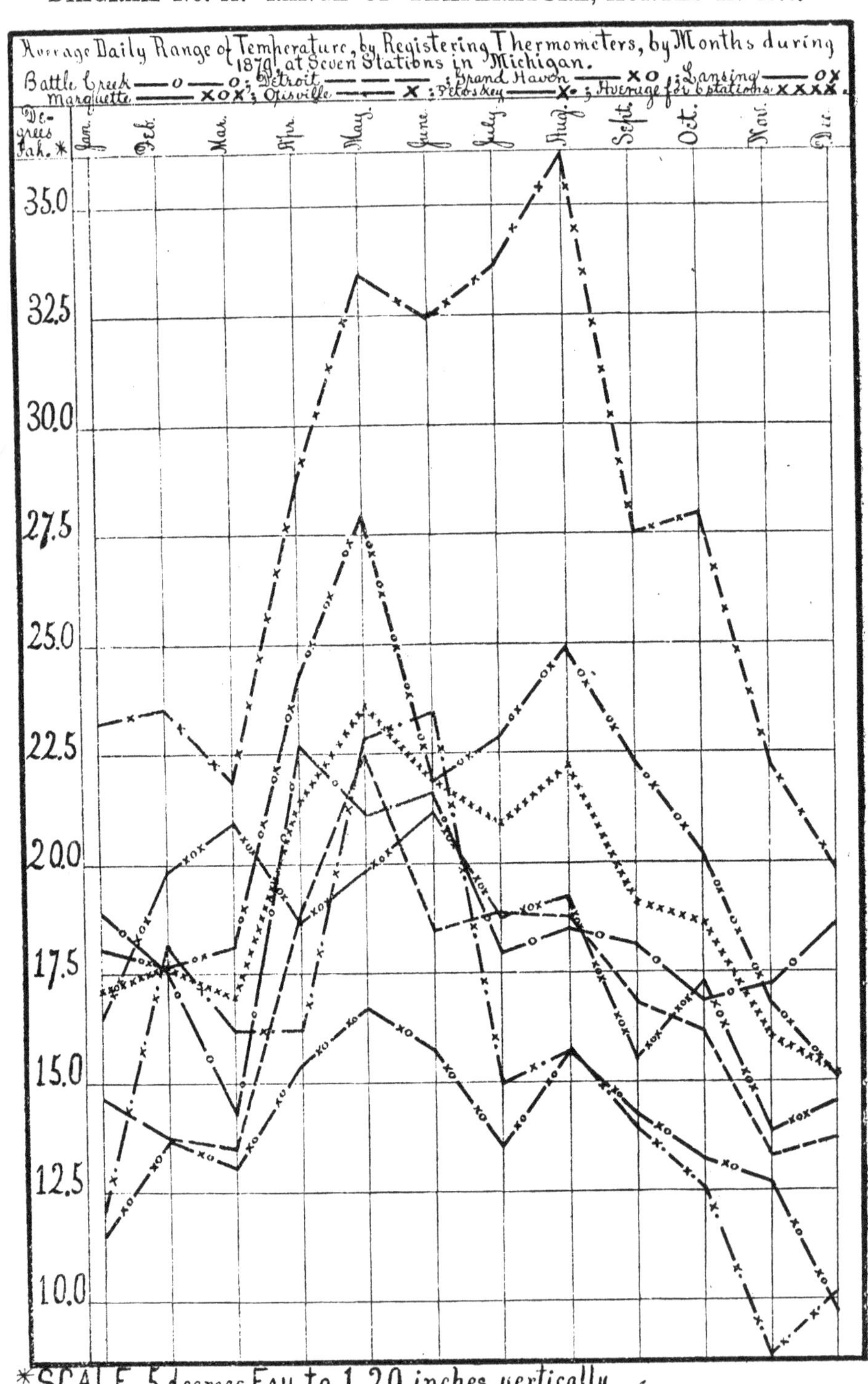

*SCALE, 5 degrees FAH. to 1.20 inches, vertically.

Designed by Henry B. Baker. Drawn by H. B. Turner and Jno. K. Allen.

EXHIBIT 14.—*Comparison of the Average Daily Range of Temperature for the Year and for each Month of the Year 1879 with Averages for the 9 Years, 1870–78, and for the Year 1878,—as indicated by Observations made at 7 A. M., 2 P. M., and 9 P. M., Daily, by Prof. R. C. Kedzie, at the State Agricultural College*, near Lansing, Michigan.*

YEARS, ETC.	AVERAGE DAILY RANGE OF TEMPERATURE,—DEGREES FAHR.												
	Annual Av.	Jan.	Feb.	Mar.	Apr.	May.	June.	July.	Aug.	Sept.	Oct.	Nov.*	Dec.*
Av. 9 Years, '70–8	15.55	11.21	15.83	14.93	16.44	16.41	17.16	16.97	17.87	19.46	16.77	13.06	10.54
1878	13.56	8.74	14.55	12.06	14.07	13.08	15.80	14.29	17.71	17.43	15.77	11.57	7.90
1879*	14.90	13.39	11.46	12.74	19.03	18.68	16.60	16.81	19.06	16.13	15.23	10.90	8.71
In 1879 **Greater** than Av. 9 Years, 1870–8	......	2.18	......	......	2.59	2.27	......	......	1.19	......	......	......	......
In 1879 **Less** than Av. 9 Years, 1870–8	.65	......	4.37	2.19	......	......	.56	.16	......	3.33	1.54	2.16	1.83
In 1879 **Greater** than in 1878	1.34	4.65	......	.68	4.96	5.60	.80	2.52	1.35	......	......	......	.81
In 1879 **Less** than in 1878	......	......	3.09	......	......	......	......	......	......	1.30	.54	.67	......

* For November and December, 1879, the observations were made by Harry B. Turner, at the office of the State Board of Health, Lansing.

TABLE II.—*Average Daily Range of Temperature, by Registering Thermometers ‡, during the Year and during each Month of the Year 1879, for Six Stations, and at each of Seven Stations in Michigan.*

STATIONS IN MICHIGAN.*	Division of the State.†	AVERAGE DAILY RANGE OF TEMPERATURE,—DEGREES FAHR.												
		Year 1879.	MONTHS, 1879.											
			Jan.	Feb.	Mar.	Apr.	May.	J'ne.	J'ly.	Aug.	Sept.	Oct.	Nov.	Dec.
Av. 6 Stations ‡	......	19.22	17.12	17.64	16.95	21.43	23.58	21.87	20.92	22.21	19.06	18.61	16.00	15.23
Marquette	U. P.†	17.99	16.48	19.82	20.94	18.60	19.84	21.17	18.71	19.16	15.47	17.29	13.83	14.52
Petoskey ‡	N.	15.40	12.03	18.14	16.19	16.20	22.84	23.47	14.97	15.71	13.90	12.58	8.67	10.13
Grand Haven	W.	13.75	11.52	13.64	13.03	15.43	16.65	15.70	13.52	15.68	14.20	13.23	12.67	9.68
Lansing	C.	20.83	18.04	17.64	18.10	24.22	27.96	21.87	22.87	24.97	22.27	20.16	16.77	15.03
Otisville	C.	27.54	23.19	23.54	21.84	28.70	33.45	32.43	33.63	36.19	27.50	27.97	22.20	19.83
Battle Creek	S. C.	18.61	18.84	17.50	14.29	22.67	21.10	21.60	17.94	18.50	18.13	16.87	17.21	18.63
Detroit	S. E.	16.60	14.65	13.71	13.48	18.93	22.48	18.43	18.84	18.74	16.80	16.13	13.30	13.68

* The names of observers, their places of observation, and the counties in which these places are situated are stated in Exhibit 7, page 306.

† For counties in each division, see Exhibit 1, page 227.

‡ The range of temperature at Petoskey, having been determined from the 7 A. M., 2 P. M., and 9 P. M. observations, has not been included in the average with the other 6 stations, for which the range was determined from registering thermometers.

Graphic representations of statements in Table II. are given in Diagram II., opposite this page.

TABLE III.—*Extremes of Temperature and Days of Month on which the Highest and for the Year 1879, at each of 23 Stations in Michigan,—as indicated by Daily Readings P. M., by Observers* for the State Board of Health, and for the U. S. Signal Service.*

Line Number.	STATIONS IN MICHIGAN.*	YEAR 1879.			JANUARY.		FEBRUARY.		MARCH.		APRIL.		MAY.	
		Highest.	Lowest.	Range.	Highest.	Lowest.	Highest.	Lowest.	Highest.	Lowest.	Highest.	Lowest.	Highest.	Lowest.
1	At 19 Stations..‡	102°	-22°	124°	50°	-20°	46°	-22°	72°	-12°	87°	5°	95°	22°
2	Marquette*.....§	96	-20	116	45[27,30]	-6[2]	39[2]	-20[27]	63[9]	-6[15,16]	80[22]	12[2]	92[29]	30[1]
3	Menominee......	----	----	----	44[27]	-12[3]	36[10]	-20[27]	58[10]	-6[3]	65[25]	14[2]	85[29]	39[7]
4	Petoskey........a	87	-16	103	46[27]	3[20]	40[18]	-16[19]	60[10]	-2[16,19]	75[23,25]	15[3]	86[19]	31[1]
5	Alpena..........§	89	-22	111	48[27]	-8[9]	40[22]	-22[27]	66[10]	-8[17]	70[27]	5[2]	86[12,30]	29[1]
6	Grand Haven...§	86	0	86	43[27]	0[2]	45[10]	2[14]	62[8,9]	5[3]	78[24]	18[3]	83[29]	30[3]
7	Nirvana.........b	94	-19	113	46[27]	-4[1]	40[10]	-19[18]	61[8]	-12[3]	87[24]	17[2]	91[13]	32[1,6]
8	Port Huron.....§	92	-14	106	42[27]	-14[3]	41[11]	-5[27]	61[9]	8[17]	73[25]	14[3]	87[30]	30[1]
9	Thornville......§	96	-11	107	49[27]	-11[3]	41[22]	-3[27]	65[10]	14[16]	80[26]	16[3]	90[13]	30[2,8]
10	Ag'l College....§	97	-18	115	44[26,27]	-18[2]	41[10]	-6[14]	66[10]	4[2]	81[24]	12[1,3]	91[13]	25[7]
11	Ionia............	----	----	----	------	------	------	--------	------	--------	------	-----	87[12,13]	36[2,3]
12	Lansing.........§	95	-20	115	48[26,27]	-20[3]	43[8]	-2[13,14]	72[10]	11[15]	83[24]	16[3]	89[13]	30[7]
13	Otisville........§	102	-13	115	50[28]	-13[3]	43[23]	-12[27]	66[11]	4[3]	82[25]	14[3]	95[14]	25[8]
14	Benton Harbor..	----	----	----	44[27,29]	-3[2,10]	48[10]	0[27]	75[8]	12[3]	87[24]	23[2,3]	------	-----
15	Niles............	96	-18	114	44[25]	-18[3]	40[8,11]	-10[27]	69[10]	-2[3]	86[23]	25[3]	87[11,18]	32[1]
16	Battle Creek..c §	91	-10	101	46[26,27]	-10[2]	42[8]	-2[26]	66[10]	11[2]	81[23]	12[2]	84[13,14]	31[7]
17	Coldwater.......	96	-18	114	45[27]	-18[3]	46[10]	-6[27]	64[10]	16[3]	82[23]	18[3]	81[30]	30[8]
18	Kalamazoo......§	93	-14	107	47[26]	-14[2]	43[10]	-5[26]	69[8]	12[15]	83[23]	14[2]	87[12,29]	28[1]
19	Mendon..........	92	-14	106	42[27]	-14[2,3]	41[8,10]	-3[14]	70[8]	14[3]	85[23]	19[3]	85[20]	36[1]
20	Tecumseh......d	92	-13	105	44[30]	-13[3]	40[11]	-4[27]	61[8,10]	15[3]	74[23,24]	17[3]	84[13,30]	34[2]
21	Ypsilanti........	90	-13	103	45[27,28]	-13[2]	42[22]	-4[27]	69[10]	15[15]	75[25,26]	22[3]	90[31]	34[7]
22	Detroit..........§	94	-15	109	44[27]	-15[3]	42[22]	-3[15,27]	67[9]	14[16]	76[25]	16[3]	87[30]	30[2]
23	Woodmere Cemetery..........§	95	-15	110	42[27]	-15[3]	42[22]	-5[27]	65[9]	12[16]	76[24]	16[3]	88[30,31]	22[8]
24	Washington.....	93	-10	103	43[27,29]	-10[2]	43[22]	1[14,27]	63[10]	15[3]	76[25]	21[2,3]	87[12]	31[1]

NOTE.—The small figures above and at the right of numbers denoting the degrees of temperature state the day or days of the month on which the highest or the lowest temperature occurred. In some cases as great an extreme occurred on other days, stated in foot-notes to the table and referred to by superior letters in the column with the names of stations.

* The names of Observers, etc., are stated in Exhibit 7, page 306.

‡ This line refers only to the 19 stations from which observations were received for every month of the year.

§ Determined by daily readings of registering thermometers made and recorded at 7 A. M. for the preceding calendar day. But at Otisville the observations were recorded for the calendar day on which they were made; and at Woodmere Cemetery the observations on the registering thermometers were made at 9 P. M. and recorded for the day on which they were made; and at the stations of the U. S. Signal Service the registering thermometers were recorded and set each day at 11 P. M. Washington time. For stations not indicated by this mark (§), the extremes were determined from the 7 A. M., 2 P. M., and 9 P. M. observations.

a At Petoskey maximum temperature of the month also on April 26.

b At Nirvana the thermometer was as low as 16° May 1.

c At Battle Creek max. also June 24 and 26.

d At Tecumseh max. also May 31.

the Lowest Temperatures occurred, by Months of the Year 1879; also Extremes and Range of Registering Thermometers or by Observations made Daily at 7 A. M., 2 P. M., and 9

JUNE.		JULY.		AUGUST.		SEPTEMBER.		OCTOBER.		NOVEMBER.		DECEMBER.		Line Number.
Highest.	Lowest.	Highest.	Lowest.	Highest.	Lowest.	Highest.	Lowest.	Highest.	Lowest.	Highest.	Lowest.	Highest.	Lowest.	
98°	30°	100°	41°	102°	35°	100°	25°	94°	20°	81°	0°	61°	-9°	1
95^{23}	32^{2}	93^{31}	46^{24}	96^{30}	37^{16}	77^{30}	35^{24}	87^{15}	22^{31}	59^{7}	8^{20}	44^{5}	$-9^{18,26}$	2
86^{23}	46^{1}	92^{14}	$62^{23, 24}$	87^{29}	52^{25}	80^{29}	40^{24}	$79^{5, 15}$	24^{31}	55^{8}	8^{21}	------	------	3
78^{15}	36^{2}	87^{3}	50^{10}	84^{2}	50^{25}	78^{11}	35^{24}	$85^{6, 16}$	28^{24}	65^{8}	11^{20}	50^{6}	-5^{30}	4
87^{9}	34^{7}	$88^{13, 15}$	48^{19}	89^{2}	42^{17}	85^{30}	30^{25}	83^{10}	24^{31}	62^{8}	0^{30}	55^{10}	-5^{26}	5
86^{24}	40^{7}	86^{15}	54^{18}	83^{2}	46^{18}	76^{1}	30^{25}	80^{11}	$31^{24,25}$	$67^{8, 9}$	14^{20}	$58^{5, 6}$	6^{26}	6
91^{24}	43^{6}	$94^{14,21}$	55^{18}	$92^{1, 30}$	40^{9}	80^{1}	25^{25}	$85^{6, 7}$	21^{25}	63^{11}	7^{24}	48^{5}	-7^{19}	7
90^{25}	39^{7}	92^{15}	53^{5}	$91^{2, 29}$	46^{10}	80^{1}	31^{25}	86^{6}	24^{31}	$65^{8, 11}$	10^{21}	$56^{6, 10}$	-2^{26}	8
94^{25}	38^{7}	96^{15}	$52^{18, 19}$	96^{31}	$46^{9, 17}$	86^{1}	30^{25}	86^{2}	26^{25}	$67^{8, 11}$	12^{21}	55^{5}	0^{26}	9
$95^{24,25}$	33^{6}	97^{14}	47^{17}	96^{31}	34^{16}	85^{1}	$27^{24,25}$	87^{6}	15^{31}	--------	--------	------	------	10
93^{24}	46^{1}	97^{15}	$62^{17, 19}$	96^{21}	48^{18}	95^{1}	33^{25}	87^{7}	22^{25}	69^{11}	$10^{3, 23}$	55^{5}	3^{26}	11
93^{25}	40^{6}	95^{15}	54^{17}	$95^{30, 31}$	41^{16}	$83^{1,30}$	30^{24}	89^{6}	22^{31}	75^{13}	13^{20}	58^{9}	-3^{25}	12
98^{26}	30^{7}	100^{15}	41^{25}	102^{1}	$35^{17,18}$	100^{1}	25^{25}	$94^{7, 12}$	20^{25}	70^{12}	3^{4}	61^{11}	-3^{27}	13
------	--------	$93^{14, 21}$	$62^{4, 18, 29}$	94^{21}	55^{9}	86^{30}	35^{25}	------	------	--------	--------	57^{5}	4^{26}	14
90^{24}	40^{6}	96^{10}	60^{18}	$90^{21, 31}$	51^{26}	83^{30}	32^{26}	88^{11}	26^{25}	81^{13}	10^{3}	59^{2}	-1^{25}	15
$89^{9, 22}$	$43^{3, 6}$	91^{15}	$55^{17, 29}$	90^{30}	42^{17}	89^{1}	$31^{21,25}$	86^{12}	28^{31}	69^{11}	$12^{21, 23}$	$58^{9, 10}$	-1^{25}	16
90^{24}	$48^{2, 3}$	92^{15}	$62^{4, 19}$	96^{31}	52^{18}	76^{30}	$45^{14,19}$	84^{12}	29^{31}	66^{11}	$16^{3, 20, 21}$	52^{6}	3^{25}	17
90^{24}	41^{6}	93^{15}	55^{18}	92^{31}	45^{16}	79^{30}	31^{24}	85^{11}	24^{31}	$69^{11, 13}$	$13^{2, 3}$	58^{9}	-3^{25}	18
87^{25}	46^{7}	92^{15}	61^{18}	92^{31}	$52^{17, 18}$	77^{30}	35^{25}	87^{12}	28^{25}	$69^{11, 13}$	8^{3}	52^{5}	0^{25}	19
$87^{24, 26}$	$40^{16,18,22}$	92^{15}	$60^{18,19,20}$	$88^{30, 31}$	$50^{17,18}$	80^{1}	33^{25}	83^{6}	27^{26}	68^{13}	$13^{21, 24}$	53^{10}	$3^{18,26}$	20
83^{27}	53^{1}	$88^{14, 15}$	$66^{24, 30}$	$88^{2,30,31}$	$57^{18,26}$	80^{1}	46^{25}	$85^{6, 12}$	30^{24}	71^{8}	14^{21}	56^{10}	0^{26}	21
91^{25}	39^{7}	94^{15}	$54^{5, 19}$	93^{30}	48^{17}	$82^{1,30}$	40^{19}	$85^{6, 7}$	$26^{25,31}$	$69^{8, 11}$	12^{21}	59^{5}	-2^{26}	22
92^{25}	37^{7}	94^{15}	$49^{1, 5}$	95^{30}	44^{17}	83^{1}	$28^{25, 26}$	$86^{1, 6,7}$	21^{25}	72^{11}	9^{21}	$59^{5, 6}$	-3^{26}	23
90^{25}	40^{6}	$91^{14, 15}$	$57^{4, 17}$	$93^{30, 31}$	52^{16}	85^{1}	33^{24}	86^{6}	$25^{26,31}$	$67^{8, 11, 12}$	8^{24}	55^{5}	3^{26}	24

EXHIBIT 15.—*Comparison of the Extremes and the Range of Temperature (Degrees Fahr.) during the Year and during each Month of the Year 1879, with the Average of the Extremes and of the Range for the Six Years, 1873–8, and also with the Extremes and Range for the Year and for each Month of the Year 1878; also, Statement of the Extremes and of the Range for each of the Six Years, 1873–8. Observations made with Registering Thermometers (except for the first two Months of 1873, and for those two Months with an Ordinary Thermometer at 7 A. M., 2 P. M., and 9 P. M.) Daily, by Prof. R. C. Kedzie, at the State Agricultural College,* near Lansing, Michigan.*

YEAR AND MONTHS.	EXTREMES AND RANGES OF TEMPERATURE,—DEGREES F.																													
	1873.			1874.			1875.			1876.			1877.			1878.			Av. for 6 Yrs., 1873–8.†			1879.*			In 1879 Higher (+), or Lower (-), than Av. for 6 Yrs, 1873–8.			In 1879 Higher, (+), or Lower (-), than in 1878.		
	Highest.	Lowest.	Range.	Highest.	Lowest.	Range.	Highest.	Lowest.	Range.	Highest.	Lowest.	Range.	Highest.	Lowest.	Range.	Highest.	Lowest.	Range.	Highest.	Lowest.	Range.	Highest.	Lowest.	Range.	Highest.	Lowest.	Range.	Highest.	Lowest.	Range.
YEAR	94	-30	124	101	-7	108	94	-33	127	96	-19	115	93	-14	107	98	-7	105	96	-18	114	97	-18	115	+1	0	+1	-1	-11	+10
Av. Month	74	15	59	77	15	62	75	10	64	74	19	56	74	20	54	73	22	51	74	17	58	76	15	61	+2	-2	+3	+3	-7	+10
January	43	-30	73	59	-7	66	35	-13	48	65	6	59	52	-9	61	48	-4	52	50	-10	60	44	-18	62	-6	-8	+2	-4	-14	+10
February	49	-13	62	48	-1	49	42	-33	75	59	-1	60	56	10	46	55	-7	62	52	-8	59	41	-6	47	-11	+2	-12	-14	+1	-15
March	57	-12	69	67	8	59	75	-11	86	60	0	60	51	-14	65	72	18	54	64	-2	66	66	4	62	+2	+6	-4	-6	+14	+8
April	82	24	58	68	3	65	80	0	80	74	16	58	81	18	63	75	29	46	77	15	62	81	12	69	+4	-3	+7	+6	+17	+23
May	84	27	57	96	21	75	89	24	65	89	31	58	90	26	64	77	29	48	88	26	61	91	25	66	+3	-1	+5	+14	+4	+18
June	94	42	52	95	34	61	89	33	56	95	42	53	89	40	49	94	39	55	93	38	54	95	33	62	+2	-5	+8	+1	+6	+7
July	92	44	48	98	43	55	92	44	48	96	46	50	91	43	48	98	47	51	95	45	50	97	47	50	+2	+2	0	-1	0	-1
August	94	44	50	101	41	60	93	35	58	96	36	60	93	43	50	93	42	51	95	40	55	96	34	62	+1	-6	+7	+3	-8	+11
September	89	26	63	95	30	65	94	26	68	80	36	44	85	38	47	92	31	61	89	31	58	85	27	58	-4	-4	0	-7	-4	-3
October	79	16	63	76	16	60	77	18	59	75	19	56	87	26	61	82	21	61	79	19	60	87	15	72	+8	-4	+12	+5	-6	+11
November *	56	1	55	70	3	67	60	2	58	62	12	50	55	4	51	52	15	37	59	6	53	75	13	62	+16	+7	+9	+23	-2	+25
December *	64	10	54	50	-6	56	70	-1	71	41	-19	60	58	13	45	36	-2	38	53	-1	54	58	-3	61	+5	-2	+7	+22	-1	+23

* For November and December, 1879, the observations were made by Harry B. Turner, at the office of the State Board of Health, Lansing.

† For the six years: highest, 101°, August 11, 1874; lowest, -33°, February 8, 1875; range, 134°.

HUMIDITY OF THE ATMOSPHERE.

That variations in the absolute humidity of the atmosphere correspond very closely with variations in temperature may be seen by a comparison of Diagram III., page 327, with Diagram I., page 316. By comparing Exhibit 16, below, with Exhibit 13, page 314, it would seem that in 1879 the humidity of the air was greater than in 1878 in every month in which the average temperature was higher; and less in every month in which the temperature was lower.

The relative humidity was greater at the College than at the office of the State Board of Health in every month. This may be in part due to difference in places of observation of temperature mentioned on page 312, the instruments at the office of the Board being placed on the south side of the State Capital.

EXHIBIT 16.—*Comparison by Year and Months of the Average Absolute Humidity (Grains of Vapor in a Cubic Foot of Air) at 16 Stations in 1879, with the Average at 12 Stations in 1878.**

YEARS, ETC.	GRAINS OF VAPOR IN A CUBIC FOOT OF AIR,—(ABSOLUTE HUMIDITY.)												
	Annual Av.	Jan.	Feb.	Mar.	April.	May.	June.	July.	Aug.	Sept.	Oct.	Nov.	Dec.
Av. for 12 Stations in 1878...	3.81	1.73	1.83	2.79	3.68	3.88	5.26	7.14	6.37	5.44	3.62	2.47	1.54
Av. for 16 Stations in 1879...	3.43	1.31	1.23	1.93	2.64	3.94	5.09	6.30	5.58	4.40	4.61	2.46	1.70
In 1879 **Greater** than in 1878....	--------	------	------	------	------	.06	------	------	------	------	.99	------	.16
In 1879 **Less** than in 1878....	.38	.42	.63	.86	1.04	------	.17	.84	.79	1.04	------	.01	------

* Ten of the stations, namely, Nirvana, Otisville, Thornville, Niles, Battle Creek, Kalamazoo, Mendon, Tecumseh, Detroit, and Woodmere Cemetery, were the same for both years; two, Benton Harbor and Coldwater, were included in the average for 1878 but not in that for 1879; and six, Marquette, Petoskey, Alpena, Grand Haven, Port Huron, and Lansing, were included in the average for 1879, but not in that for 1878.

TABLE IV.—*The Average Number of Grains of Vapor of Water in a Cubic Foot of Air (Absolute Humidity), for Months and Year 1879, at 22 Stations in Michigan,—Average of Observations made Daily at 7 A. M., 2 P. M., and 9 P. M., by Observers* for the State Board of Health, and for the U. S. Signal Service.*

STATIONS IN MICHIGAN.*	GEOGRAPHICAL DIVISIONS OF THE STATE.†	GRAINS OF VAPOR IN A CUBIC FOOT OF AIR.													
		YEAR.		MONTHS, 1879.											
		1878.	1879.	Jan.	Feb.	Mar.	Apr.	May.	June.	July.	Aug.	Sept.	Oct.	Nov.	Dec.
Av. for 16 Stations‡	---------	[a]	3.43	1.31	1.23	1.93	2.64	3.94	5.09	6.30	5.58	4.40	4.61	2.46	1.70
Marquette........*	U. P. †	-----	2.90	1.08	0.94	1.47	2.15	3.23	4.46	5.51	5.03	3.88	3.90	1.94	1.18
Menominee........	U. P.	-----	[b]	------	-----	-----	-----	3.71	4.95	6.30	5.58	4.23	4.26	[h]1.92	-----
Petoskey..........	N.	-----	3.40	1.58	1.25	2.02	2.45	3.50	4.42	5.26	6.08	5.21	4.74	2.54	1.70
Alpena............	N. E.	-----	2.93	1.02	0.84	1.28	1.98	3.15	4.39	5.76	5.19	4.04	4.10	2.10	1.34
Grand Haven......	W.	-----	3.48	1.33	1.22	2.00	2.57	3.97	4.91	6.40	5.77	4.41	4.78	2.65	1.78
Nirvana...........	W.	3.40	3.13	1.24	1.00	1.61	[i]2.23	3.69	4.84	5.91	5.08	4.02	4.38	2.10	1.42
Port Huron........	B. & E.	-----	3.23	1.15	1.15	1.82	2.39	3.47	4.79	6.02	5.30	4.18	4.38	2.39	1.74
Thornville	B. & E.	3.87	3.76	1.51	1.44	2.15	2.80	4.38	5.67	6.85	5.69	4.68	5.09	2.82	2.02
Ionia..............	C.	-----	[c]	------	-----	-----	-----	4.17	5.25	6.43	5.41	4.46	4.54	-----	-----
Lansing...........	C.	-----	3.44	[j]1.48	1.40	2.07	[k]2.75	[l]4.19	5.18	6.10	5.26	4.19	4.58	2.45	1.65
Otisville..........	C.	3.76	3.59	1.45	1.36	2.00	2.73	3.99	5.16	6.37	6.00	4.75	4.88	2.53	1.85
Benton Harbor....	S. W.	4.80	[d]	[m]1.82	1.99	2.87	[n]4.08	-----	------	8.84	8.28	[o]6.20	-----	[p]2.59	2.26
Niles..............	S. W.	4.02	3.88	1.21	1.27	2.20	3.61	4.96	6.02	7.20	6.08	4.60	4.88	2.77	1.81
Battle Creek......	S. C.	3.67	3.67	1.49	1.41	2.19	3.23	4.22	5.46	6.85	5.73	4.39	4.77	2.48	[q]1.76
Coldwater.........	S. C.	3.94	[e]	------	1.02	2.04	2.81	3.68	[r]6.47	[s]8.09	5.84	4.20	[m]4.60	-----	1.62
Kalamazoo.........	S. C.	3.62	3.44	1.32	1.23	1.95	2.55	3.99	5.17	6.37	5.37	4.27	4.70	2.50	1.81
Mendon............	S. C.	3.78	3.60	[t]1.44	1.46	[t]2.04	2.79	4.23	5.48	6.62	5.54	4.44	4.73	2.59	1.85
Tecumseh	S. C.	3.87	3.47	1.29	1.36	2.12	2.82	4.06	4.65	6.58	5.68	4.44	4.57	2.44	1.61
Ypsilanti...	S. C.	-----	[f]	------	-----	-----	-----	[u]4.65	5.74	6.36	[v]5.76	4.61	4.64	2.38	[i]1.73
Detroit.............	S. E.	3.34	3.48	1.24	1.15	1.98	2.60	3.94	5.23	6.32	5.68	4.42	4.71	2.59	1.88
Woodmere Cemetery, near Detroit	S. E.	3.68	3.51	1.18	1.19	1.93	2.61	4.02	5.53	6.61	5.73	4.44	4.63	2.47	1.80
Washington.......	S. E.	-----	[g]	------	-----	-----	2.60	3.82	5.21	6.31	5.17	4.23	4.60	2.41	1.78

* The names of observers, their places of observation, and the counties in which these places are situated, are stated in Exhibit 7, page 306.

† The full names of the divisions and the counties in each division, are stated in Exhibit 1, page 227.

‡ This line is an average for only the stations for which statements are given for every month of the year.

[a] The average for 12 stations in 1878 is 3.81 grs. [b] For 7 months, 4.42. [c] For 6 months, 5.04. [d] For 9 months, 4.33. [e] For 10 months, 4.04. [f] For 8 months, 4.48. [g] For 9 months, 4.01. [h] For about 80 observations. [i] For 22 days. [j] For the last 25 days. [k] For 84 observations. [l] For 85 observations. [m] For 86 observations. [n] For 81 observations. [o] For 70 observations. [p] For last 15 days. [q] For 29 days. [r] For 69 observations. [s] For 77 observations. [t] For 87 observations. [u] For 30 days. [v] For 83 observations.

The lines for 11 representative stations in Table IV., are graphically represented in Diagram III., opposite this page. Comments are printed on pages 307–308.

DIAGRAM NO. III.—ABSOLUTE HUMIDITY, BY MONTHS, IN 1879.

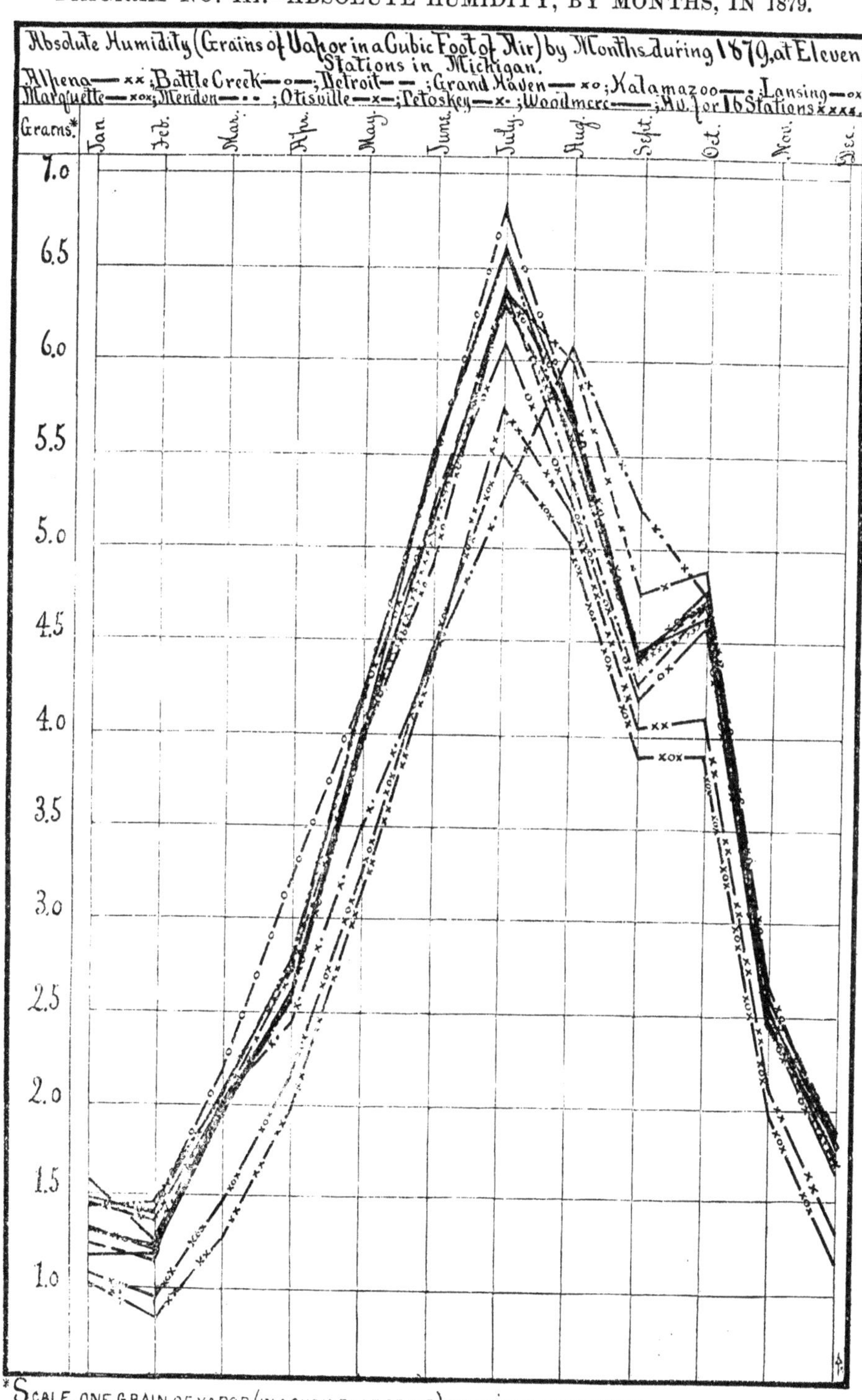

*SCALE, ONE GRAIN OF VAPOR (IN A CUBIC FOOT OF AIR) TO AN INCH, VERTICALLY.

Drawn by H. B. Turner, and Jno K. Allen. Designed by Henry B. Baker.

DIAGRAM No. IV.--RELATIVE HUMIDITY, BY MONTHS IN 1879.

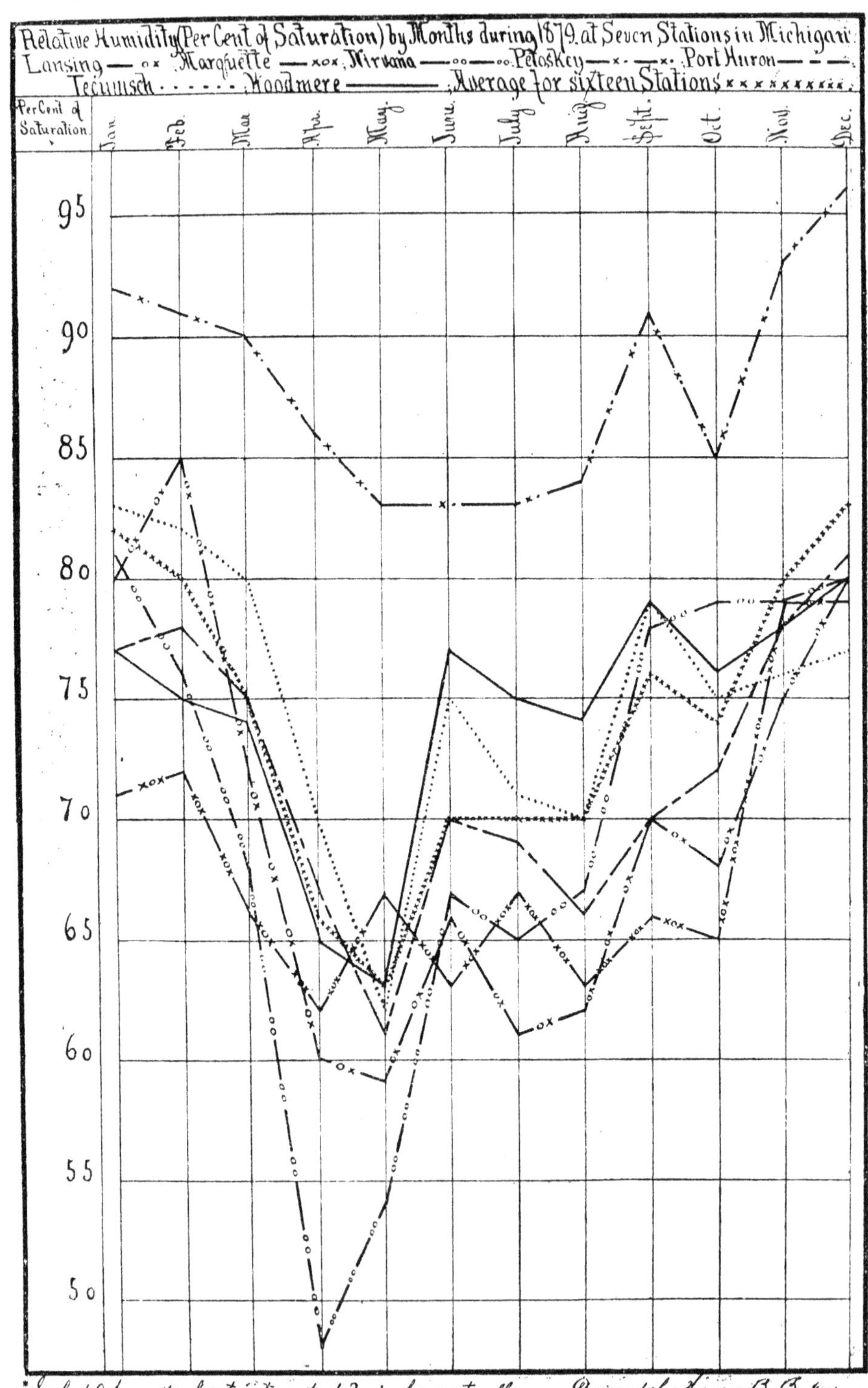

*Scale, 10 per cent of saturation to 1.3 inches, vertically. Designed by Henry B. Baker.

Drawn by H. B. Turner and Jno. H. Allen.

TABLE V.—*Average Per Cent of Saturation of the Atmosphere with Vapor of Water (Relative Humidity) during the Year and during each Month of the Year 1879, at 23 Stations in Michigan,—Average of Observations made Daily at 7 A. M., 2 P. M., and 9 P. M., by Observers* for the State Board of Health, and for the U. S. Signal Service.*

STATIONS IN MICHIGAN.*	Geographical Divisions of the State.†	PER CENT OF SATURATION,—RELATIVE HUMIDITY.													
		YEAR.		MONTHS, 1879.											
		1878.	1879.	Jan.	Feb.	Mar.	Apr.	May.	J'ne	J'ly.	Aug.	Sept.	Oct.	Nov.	Dec.
Av. for 16 Stations‡	....	a	74	82	80	75	66	63	70	70	70	76	74	80	83
Marquette........*	U. P.†	.....	68	71	72	66	62	67	63	67	63	66	65	79	79
Menominee........	U. P.	.....	b	.....	.....	.....	.....	63	63	66	66	69	68	i 70	.....
Petoskey..........	N.	.....	88	92	91	90	86	83	83	83	84	91	85	93	96
Alpena............	N. E.	.....	72	71	71	68	64	59	68	70	76	73	75	82	81
Grand Haven......	W.	.....	72	74	75	72	64	63	66	70	70	74	77	81	81
Nirvana...........	W.	74	70	81	76	68	j 48	54	67	65	67	78	79	79	80
Port Huron........	B. & E.	.....	72	77	78	75	67	61	70	69	66	70	72	78	81
Thornville.........	B. & E.	81	79	92	91	82	68	66	74	71	67	80	80	88	93
Ag. College........	C.	78	c	89	92	78	66	64	75	71	70	78	76	.....	.....
Ionia..............	C.	.....	d	.....	.....	.....	.....	63	69	65	62	75	77	.....	.....
Lansing...........	C.	.....	70	k 80	85	72	l 60	m 59	66	61	62	70	68	75	80
Otisville..........	C.	81	78	92	89	78	67	59	70	71	77	82	71	86	94
Benton Harbor....	S. W.	.....	e	n 99	100	100	o 97	.....	.....	94	94	p 98	.....	q 98	100
Niles..............	S. W.	79	77	73	76	80	78	75	79	78	76	79	73	82	71
Battle Creek......	S. C.	69	70	88	82	73	68	59	61	66	61	68	69	67	r 78
Coldwater.........	S. C.	78	f	.....	60	68	61	57	s 82	t 85	69	71	n 68	.....	79
Kalamazoo........	S. C.	75	72	84	79	71	59	60	67	66	63	73	75	80	88
Mendon...........	S. C.	78	78	u 93	88	u 76	66	66	73	72	70	79	76	84	87
Tecumseh.........	S. C.	80	75	83	82	80	70	62	75	71	70	79	75	76	77
Ypsilanti..........	S. C.	.....	g	.....	.....	.....	.....	v 64	74	68	w 65	69	64	66	i 77
Detroit............	S. E.	71	70	76	71	71	61	59	68	68	67	72	73	77	82
Woodmere Cemetery.	S. E.	77	74	77	75	74	65	63	77	75	74	79	76	78	80
Washington........	S. E.	.....	h	.....	.....	.....	67	63	75	70	66	78	76	82	86

* The names of observers, their places of observation, and the counties in which these places are situated, are stated in Exhibit 7, page 306.

† The full names of the divisions and the counties in each division, are stated in Exhibit 1, page 227.

‡ This line is an average for only the stations for which statements are given for every month of the year.

[a] The average for 12 stations in 1878 is 78. [b] For 7 months, 66. [c] Including the observations at Lansing for Nov. and Dec., the average of this line is 76. [d] For 6 months, 69. [e] For 9 months, 98. [f] For 10 months, 70. [g] For 8 months, 68. [h] For 9 months, 74. [i] For about 80 observations. [j] For 22 days. [k] For the last 25 days. [l] For 84 observations. [m] For 85 observations. [n] For 86 observations. [o] For 81 observations. [p] For 70 observations. [q] For last 15 days. [r] For 29 days. [s] For 69 observations. [t] For 77 observations. [u] For 87 observations. [v] For 30 days. [w] For 83 observations.

Graphic representations of 7 representative lines in this table are given in Diagram IV., opposite this page.

EXHIBIT 17.—*Comparison of the Average Relative Humidity of the Air (Per Cent of Saturation) for the Year and for each Month of the Year 1879, with Averages for the 15 Years, 1864–78, and for 1878.—Observations made at 7 A. M., 2 P. M., and 9 P. M., Daily, by Prof. R. C. Kedzie, at the State Agricultural College*, near Lansing, Michigan.*

YEARS, ETC.	PER CENT OF SATURATION.												
	Annual Av.	Jan.	Feb.	Mar.	Apr.	May.	June.	July.	Aug.	Sept.	Oct.	Nov.*	Dec.*
Average 15 Y'rs, 1864–78	79	86	86	85	71	69	76	76	77	81	79	83	86
1878	78	89	89	83	69	61	73	68	74	78	76	87	95
1879*	76	89	92	78	66	64	75	71	70	78	76	75	80
In 1879 **Greater** than Av. 15 Years, 1864–78	-------	3	6	------	------	------	------	------	------	------	------	------	------
In 1879 **Less** than Av. 15 Years, 1864–78	3	------	------	7	5	5	1	5	7	3	3	8	6
In 1879 **Greater** than in 1878	-------	------	3	------	------	3	2	3	------	------	------	------	------
In 1879 **Less** than in 1878	2	------	------	5	3	------	------	------	4	------	------	12	15

* For November and December, 1879, the observations were made by Harry B. Turner, at the office of the State Board of Health, Lansing.

CLOUDINESS.

EXHIBIT 18.—*Comparison of the Average Per Cent of Cloudiness in the Year and in each Month of the Year 1879, with Averages for the 15 Years 1864–78, and for the Year 1878. Observations made at 7 A. M., 2 P. M., and 9 P. M., Daily, by Prof. R. C. Kedzie, at the State Agricultural College,* near Lansing, Michigan.*

YEARS, ETC.	PER CENT OF CLOUDINESS.												
	Annual Av.	Jan.	Feb.	Mar.	Apr.	May.	June.	July.	Aug.	Sept.	Oct.	Nov.*	Dec.*
Av. 15 Years, 1864–78	59	74	63	65	58	52	50	47	48	50	59	68	76
1878	58	76	56	72	59	54	45	37	40	45	54	68	89
1879*	54	68	73	56	44	40	43	36	35	47	55	68	84
In 1879 **Greater** than Av. 15 Years, 1864–78	-------	----	10	------	----	------	------	------	------	------	----	0	8
In 1879 **Less** than Av. 15 Years, 1864–78	5	6	----	9	14	12	7	11	13	3	4	0	------
In 1879 **Greater** than in 1878	-------	----	17	------	----	------	------	------	------	2	1	0	------
In 1879 **Less** than in 1878	4	8	----	16	15	14	2	1	5	------	----	0	5

* For November and December, 1879, the observations were made by Harry B. Turner, at the office of the State Board of Health, Lansing.

DIAGRAM No. V.—CLOUDINESS BY MONTHS IN 1879.

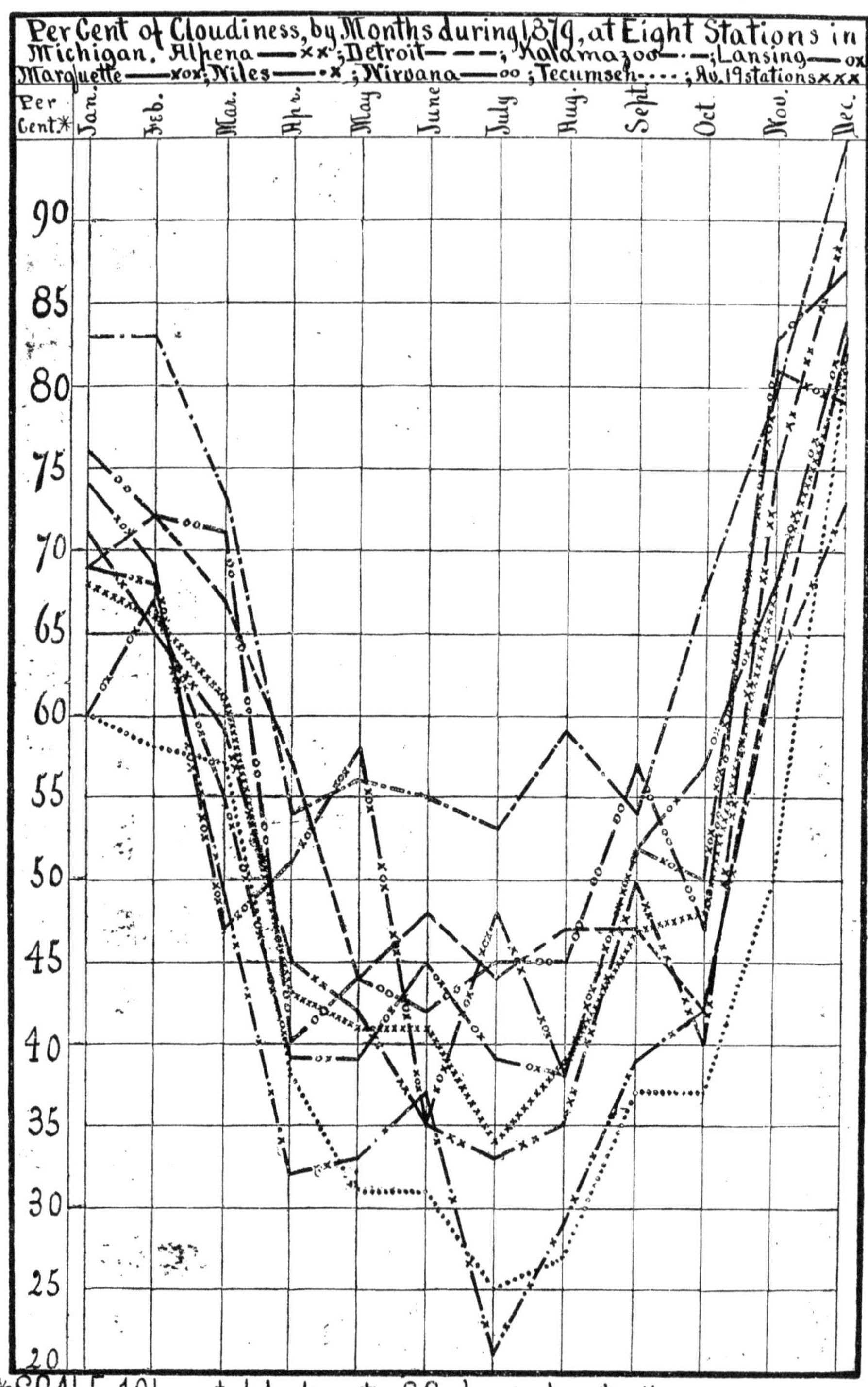

*SCALE, 10 per cent of cloudiness to .88 of an inch, vertically.

Designed by Henry B. Baker. Drawn by H. B. Turner and Jno K. Allen.

TABLE VI.—*Average Per Cent of Cloudiness, for the Year and for each Month of the Year 1879, at 23 Stations in Michigan,—Average of Observations made Daily at 7 A. M., 2 P. M., and 9 P. M., by Observers for the State Board of Health*, and for the U. S. Signal Service.*

STATIONS IN MICHIGAN.*	Geographical Divisions of the State.†	AVERAGE PER CENT OF CLOUDINESS.													
		YEAR.		MONTHS, 1879.											
		1878.	1879.	Jan.	Feb.	Mar.	Apr.	May.	J'ne.	J'ly.	Aug.	Sept.	Oct.	Nov.	Dec.
Av. for 23 Stations‡	----------	[a]	53	68	66	61	43	41	41	34	39	47	48	68	82
Marquette--------*	U. P.†	-----	57	74	69	47	51	58	35	48	38	52	50	81	79
Menominee--------	U. P.	-----	[b]	43	49	53	37	51	33	39	41	51	44	62	-----
Petoskey----------	N.	55	52	69	46	63	37	35	38	31	25	58	51	90	81
Alpena ------------	N. E.	-----	53	71	65	59	45	42	35	33	35	50	40	75	90
Grand Haven------	W.	-----	54	77	73	61	34	42	40	25	37	51	51	77	82
Nirvana-----------	W.	59	59	76	72	71	40	44	42	45	45	57	47	83	87
Port Huron--------	B. & E.	-----	58	73	69	63	57	49	48	45	49	50	47	64	85
Thornville --------	B. & E.	55	52	64	69	57	48	36	45	34	37	43	44	64	81
Agrl. College------	C.	58	[c]	68	73	56	44	40	43	36	35	47	55	-----	-----
Ionia --------------	C.	-----	[d]	-----	-----	-----	-----	33	43	45	50	53	52	65	77
Lansing -----------	C.	-----	54	[f] 60	67	55	39	[g] 39	45	39	38	52	57	68	84
Otisville-----------	C.	60	51	58	64	58	45	39	46	35	44	49	41	61	77
Benton Harbor----	S. W.	50	[e]	72	68	57	32	-----	-----	21	22	56	-----	73	84
Niles---------------	S. W.	48	46	69	68	49	32	33	37	21	29	39	42	63	73
Battle Creek------	S. C.	51	50	78	72	64	39	37	37	27	26	39	45	64	[h] 77
Coldwater---------	S. C.	45	50	63	61	[i] 64	23	[h] 45	42	27	36	43	53	60	86
Kalamazoo--------	S. C.	67	68	83	83	73	54	56	55	53	59	54	68	80	95
Mendon -----------	S. C.	52	50	[j] 58	62	[g] 57	37	39	40	32	36	39	55	71	78
Tecumseh---------	S. C.	50	44	60	58	57	38	31	31	25	27	37	37	50	81
Ypsilanti ---------	S. C.	-----	53	[k] 68	62	66	50	36	42	28	43	44	53	64	[l] 83
Detroit -----------	S. E.	-----	57	69	72	67	57	44	48	44	47	47	42	64	83
Woodmere Cemetery.	S. E.	51	49	57	66	61	47	36	38	28	41	39	41	55	75
Washington-------	S. E.	-----	52	65	[m] 65	65	50	42	42	29	42	43	39	61	79

* The names of observers, their places of observation, and the counties in which these places are situated, are stated in Exhibit 7, page 306. The average per cent of cloudiness as observed at the Agricultural College, was greater than at the office of the State Board of Health from January to May inclusive, and less the remaining months of the year. Inasmuch as the estimation of the per cent of cloudiness may be differently made by different persons, it may be best to state that about June 1 there was a change of observers at the office of the State Board of Health.

† The full names of divisions and the counties in each division are stated in Exhibit 1, page 227.

‡ This line is an average for only the stations from which statements were received for every month of the year.

[a] The average for 13 stations in 1878 is 54 per cent. [b] For 11 months, 46. [c] Including the observations at Lansing for Nov. and Dec., the average of this line is 54 per cent. [d] For 8 months, 52 [e] For 9 months, 54. [f] For the last 25 days. [g] For 86 observations. [h] For last 29 days. [i] For first 17 days. [j] For 87 observations. [k] For 30 days. [l] For 28 days. [m] For 27 days.

Graphic representations of 8 representative lines in this table are given in Diagram V., opposite this page.

EXHIBIT A.—*Number of Days on which Fog was Recorded in 1879, and in each Month in 1879; and Dates and Hours of Observation when Fogs were Recorded, at Stations Named.*

STATIONS IN MICHIGAN.*	No. of Days, in 1879.	JANUARY.		FEBRUARY.		MARCH.		APRIL.	
		No. of Days and Day of Month.	Hour of Observation.	No. of Days and Day of Month.	Hour of Observation.	No. of Days and Day of Month.	Hour of Observation.	No. of Days and Day of Month.	Hour of Observation.
NUMBER OF DAYS, IN THE STATE		6		3		1		6	
Menominee*	2								
Alpena	9							25 26	9 P. M. 7 A. M. & 2 P. M.
Grand Haven	3	27	7 A. M.						
Nirvana	10					29	7 A. M.	25 26	6 P. M. to 6 A. M.
Port Huron	8	12 27	7 A. M. 9 P. M.						
Thornville	8	12 27	8 to 9 A. M. P. M. till sunset.						
Ionia	59							16, 18, 22, & 28	Morn.
Lansing	12	27	2 P. M.						
Otisville	9								
Niles	1								
Battle Creek	1								
Coldwater	1								
Mendon	3							16	Morn.
Tecumseh	1	27	P. M.						
Detroit	17			8	7 A. M.				
Washington	10	27	9 P. M.						
Woodmere Cemetery	34	11 13 14 and 30 27	Eve. Morn & Eve. Morn. 9 P. M.	7 and 8 10	Morn. Morn.			26	To 6 A. M.

* The names of obervers are given in Exhibit 7, page 306.

It is probable that some observers are more careful than others to note and record fogs, and that therefore Exhibit A is somewhat incomplete. By it, however, it is possible to trace the course of some general fogs, as that of January 27, which seems on that day to have swept across the State from west to east.

EXHIBIT A.—CONTINUED.—*Number of Days, and Dates when Fogs were Recorded in 1879.*

STATIONS IN MICHIGAN.*	MAY.		JUNE.		JULY.		AUGUST.	
	No. of Days and Day of Month.	Hour of Observation.	No. of Days and Day of Month.	Hour of Observation.	No. of Days and Day of Month.	Hour of Observation.	No. of Days and Day of Month.	Hour of Observation.
NUMBER OF DAYS, IN THE STATE	2		4		11		23	
Menominee			26	7 A. M.			11 and 19	7 A. M.
Alpena	14	7 A. M.						
Grand Haven							5, 7, 8, 9, 14, & 23	Morn.
Nirvana							14	7 A. M.
Port Huron								
Thornville	16	Morn.					14	Morn.
Ionia					10, 12, 14, 18, 21, and 28-30.	Morn.	2, 4, 6-9, 11, 14, 17-20, 22, 23, 25-31	Morn.
Lansing	16	7 A. M.	22	7 A. M.			4 and 14	7 A. M.
Otisville					12	7 A. M.	6 and 14	7 A. M.
Niles								
Battle Creek							19	5 to 7 A. M.
Coldwater								
Mendon								
Tecumseh								
Detroit								
Washington			18, 23, and 26	Morn.	23; 12, 13, and 23	7 A. M.; Morn.	1 and 6	Morn.
Woodmere Cemetery								

* The names of observers are given in Exhibit 7, page 306.

EXHIBIT A.—Continued.—*Number of Days, and Dates when Fogs were Recorded in 1879.*

STATIONS IN MICHIGAN.*	September.		October.		November.		December.	
	No. of Days and Day of Month.	Hour of Observation.	No. of Days and Day of Month.	Hour of Observation.	No. of Days and Day of Month.	Hour of Observation.	No. of Days and Day of Month.	Hour of Observation.
Number of Days, in the State	18	-----	21	-----	9	-----	3	-----
Menominee*	-----	-----	1	7 A. M.	-----	-----	-----	-----
Alpena	-----	-----	8 and 9	Morn., P. M., Eve.	14	7 A. M.	-----	-----
	-----	-----	16	7 A. M.	-----	-----	-----	-----
Grand Haven	-----	-----	9 and 15	7 A. M.	-----	-----	-----	-----
Nirvana	-----	-----	10	Till 8 A. M.	-----	-----	-----	-----
	-----	-----	-----		-----	-----	-----	-----
Port Huron	-----	-----	11	7 A. M., 2 P. M., and 9 P. M.	-----	-----	-----	-----
	-----	-----	12	7 A. M. & 9 P. M.	-----	-----	-----	-----
	-----	-----	15, 16, and 21	7 A. M.	10 and 11	Morn.	5	About sunset.
Thornville	-----	-----	-----	-----	13	Morn.	-----	-----
Ionia	1, 5, 10, 11, 13, 14, 17-21, 23, 29, 30	Morn.	1, 3, 4, 10–12, 14–17, and 19	Morn.	15	Morn.	-----	-----
	10 and 23	Morn.	4, 13, and 15	Morn.	13	To 9 A. M.	-----	-----
Lansing	-----	-----	14	Morn. & day.	-----	-----	-----	-----
	23	7 A. M.	13 and 15	7 A. M.	8	9 P. M.	-----	-----
Otisville	-----	-----	-----	-----	10 and 13	7 A. M.	-----	-----
	-----	-----	22	7 A. M.	-----	-----	-----	-----
Niles	-----	-----	-----	-----	-----	-----	-----	-----
Battle Creek	-----	-----	-----	-----	15	7 A. M.	-----	-----
Coldwater	-----	-----	14 and 15	7 A. M.	-----	-----	-----	-----
Mendon	-----	-----	-----	-----	-----	-----	-----	-----
Tecumseh	-----	-----	11, 13–16, 21, 25	7 A. M.	6 and 21	7 A. M.	2	9 P. M.
Detroit	-----	-----	14, 15, 19, & 24	9 P. M.	-----	-----	3	2 P. M., & 9. P. M.
	-----	-----	-----	-----	-----	-----	5	7 A. M.
Washington	-----	-----	15, 16, and 22	7 A. M.	13 and 27	7 A. M.	3	Morn.
	1, 10, 23, 25, 26	Morn.	6, 7, 10, 11, 16, 20, 21	Morn.	6	9 P. M.	-----	-----
	22	Morn & 9 P. M.	9	After 8 P. M.	10	7 A. M.	-----	-----
	24	9 P. M.	13	7 A. M., nearly all day & Eve.	13	Morn.	-----	-----
Woodmere Cemetery	-----	-----	14	To 11 A. M., & in Eve.	-----	-----	-----	-----
	-----	-----	15	To 9 A. M., & after 6 P. M.	-----	-----	-----	-----

* The names of observers are given in Exhibit 7, page 306.

RAINFALL.

At the Agricultural College the rainfall in 1879 was 5.17 inches less than that in 1878, being less in every month except November and December; it was also 3.62 inches less than the average rainfall for the preceding 15 years, being less in every month except September, November and December.

EXHIBIT 19.—*Comparison of the Rainfall during the Year and during each Month of the Year 1879, with that for the Year 1878, and with the Average for the 15 Years 1864–78. Observations made by Prof. R. C. Kedzie, at the State Agricultural College,* near Lansing, Michigan.*

YEARS, ETC.	INCHES OF RAIN AND MELTED SNOW.												
	Annual Av.	Jan.	Feb.	Mar.	April.	May.	June.	July.	Aug.	Sept.	Oct.	Nov.*	Dec.*
Av. 15 Years, 1864–78	30.44	1.76	1.70	2.83	2.49	2.85	3.87	3.33	2.64	2.99	2.31	1.88	1.83
1878	31.99	1.12	2.74	3.12	3.76	3.44	3.15	2.96	1.85	3.43	1.99	2.16	2.27
1879*	26.82	.49	1.43	1.57	1.25	2.45	2.87	2.19	1.61	3.19	1.57	4.55	3.56
In 1879 **Greater** than Av. 15 Years, 1864-78	------	----	----	------	------	------	------	------	------	.20	----	2.67	1.73
In 1879 **Less** than Av. 15 Years, 1864-78	3.62	1.27	.27	1.26	1.24	.40	1.00	1.14	1.03	------	.74	------	------
In 1879 **Greater** than in 1878	------	----	----	------	------	------	------	------	------	------	----	2.39	1.29
In 1879 **Less** than in 1878	5.17	.63	1.31	1.55	2.51	.99	.28	.77	.24	.24	.42	------	------

* For November and December, 1879, the observations were made by Harry B. Turner, at the office of the State Board of Health, Lansing.

TABLE VII.—*Inches of Rain and Melted Snow, for the Year and for each Month of the Year 1879, at 12 Stations in Michigan,—as compiled from Daily Observations made by Observers* for the State Board of Health, and for the U. S. Signal Service.*

STATIONS IN MICHIGAN.*	Geographical Divisions of the State.†	INCHES OF RAIN AND MELTED SNOW.													
		YEAR.		MONTHS, 1879.											
		1878.	1879.	Jan.	Feb.	Mar.	Apr.	May.	June.	July.	Aug.	Sept.	Oct.	Nov.	Dec.
Av. for 12 Stations‡	----------	a	36.38	1.16	2.26	1.94	1.74	2.28	3.31	3.76	2.17	5.97	2.31	5.37	4.11
Marquette.........*	U. P.†	-----	40.48	1.41	3.75	1.98	0.53	3.19	3.32	4.29	2.43	3.62	4.28	5.43	6.25
Petoskey	N.	-----	b	-----	-----	-----	0.87	4.01	1.88	3.02	4.08	5.72	5.58	-----	-----
Alpena.............	N. E.	-----	39.94	1.63	3.17	1.05	2.19	1.74	3.67	3.77	3.71	8.00	1.92	5.35	3.74
Grand Haven.......	W.	-----	35.35	1.81	1.86	1.19	1.49	2.46	2.42	2.61	2.47	6.01	2.58	7.23	3.22
Nirvana.............	W.	39.18	34.95	0.88	2.65	2.10	1.61	1.23	2.96	0.96	2.83	5.44	4.83	6.80	2.66
Port Huron.........	B. & E.	-----	27.23	0.84	1.27	1.56	1.78	1.32	3.25	1.97	0.64	4.37	1.42	4.84	3.97
Thornville...........	B. & E.	35.24	31.50	0.56	1.44	1.75	1.95	3.14	1.78	2.55	1.49	5.25	2.12	4.82	4.65
Agrl. College.......	C.	31.99	c	0.49	1.43	1.57	1.25	2.45	2.87	2.19	1.61	3.19	1.57	-----	-----
Ionia..................	C.	-----	d	-----	-----	-----	-----	3.40	2.40	3.65	1.35	5.95	5.90	4.06	2.57
Lansing	C.	-----	e	-----	-----	-----	-----	-----	-----	1.98	1.39	4.18	2.34	4.55	3.56
Otisville.............	C.	37.63	29.52	0.49	0.77	1.57	1.95	2.86	1.88	1.80	3.70	6.22	1.21	3.37	3.70
Niles..................	S. W.	38.06	45.84	0.60	4.90	3.05	2.58	1.38	5.72	5.40	1.32	7.47	2.60	6.44	4.38
Battle Creek......	S. C.	25.40	f	-----	-----	-----	-----	-----	2.50	1.60	1.40	2.22	1.53	-----	-----
Coldwater..........	S. C.	40.24	42.96	2.35	1.50	2.80	1.78	2.83	4.27	4.70	2.37	6.63	2.10	6.36	5.27
Kalamazoo	S. C.	45.82	37.53	1.10	1.39	2.69	2.23	2.31	3.12	6.62	0.84	6.27	2.91	4.49	3.56
Mendon..............	S. C.	36.87	g	-----	1.13	-----	1.66	2.88	4.40	4.96	0.59	7.72	1.95	6.10	4.06
Tecumseh	S. C.	39.89	h	-----	2.16	2.78	2.83	4.04	3.69	2.97	2.13	5.01	1.10	5.50	3.84
Ypsilanti............	S. C.	-----	i	-----	2.05	-----	1.22	1.22	-----	-----	-----	4.47	0.95	-----	2.80
Detroit	S. E.	43.39	37.17	1.11	2.64	1.68	1.56	2.64	3.85	6.22	1.31	6.23	0.83	4.71	4.39
Woodmere Cemetery.	S. E.	42.21	34.12	1.09	1.82	1.87	1.21	2.23	3.47	4.28	2.90	6.13	0.95	4.60	3.57
Washington	S. E.	-----	j	-----	-----	1.60	1.73	1.97	4.31	4.41	0.43	5.12	2.29	5.01	3.45

* The names of observers, their places of observation, and the counties in which these places are situated, are given in Exhibit 7, page 306.

† The names of the divisions, and the counties in each, are stated in Exhibit 1, page 227.

‡ This line is an average for only the 12 stations from which statements were received for every month of the year.

a The average for 12 stations in 1878 is 37.99 in. b For 7 months, 25.16 in. c Including the observations at Lansing for Nov. and Dec., the total for the year is 26.82 in. d For 8 months, 29.28 in. e For 6 months, 18.00 in. f For 5 months, 9.25 in. g For 10 months, 35.45 in. h For 11 months, 36.05 in. i For 6 months, 12.71 in. j For 10 months, 30.32 in.

The lines for 8 representative stations in Table VII. are graphically represented in Diagram VI., opposite this page.

OZONE.

At the Agricultural College in 1879 the day ozone was less than the average for the preceding seven years for the year and in every month except July and September; and the night ozone was less than the average for the same period for the year and in every month except August and September. The variation from 1878 was not so uniform.

DIAGRAM No. VI.—RAINFALL, BY MONTHS IN 1879.

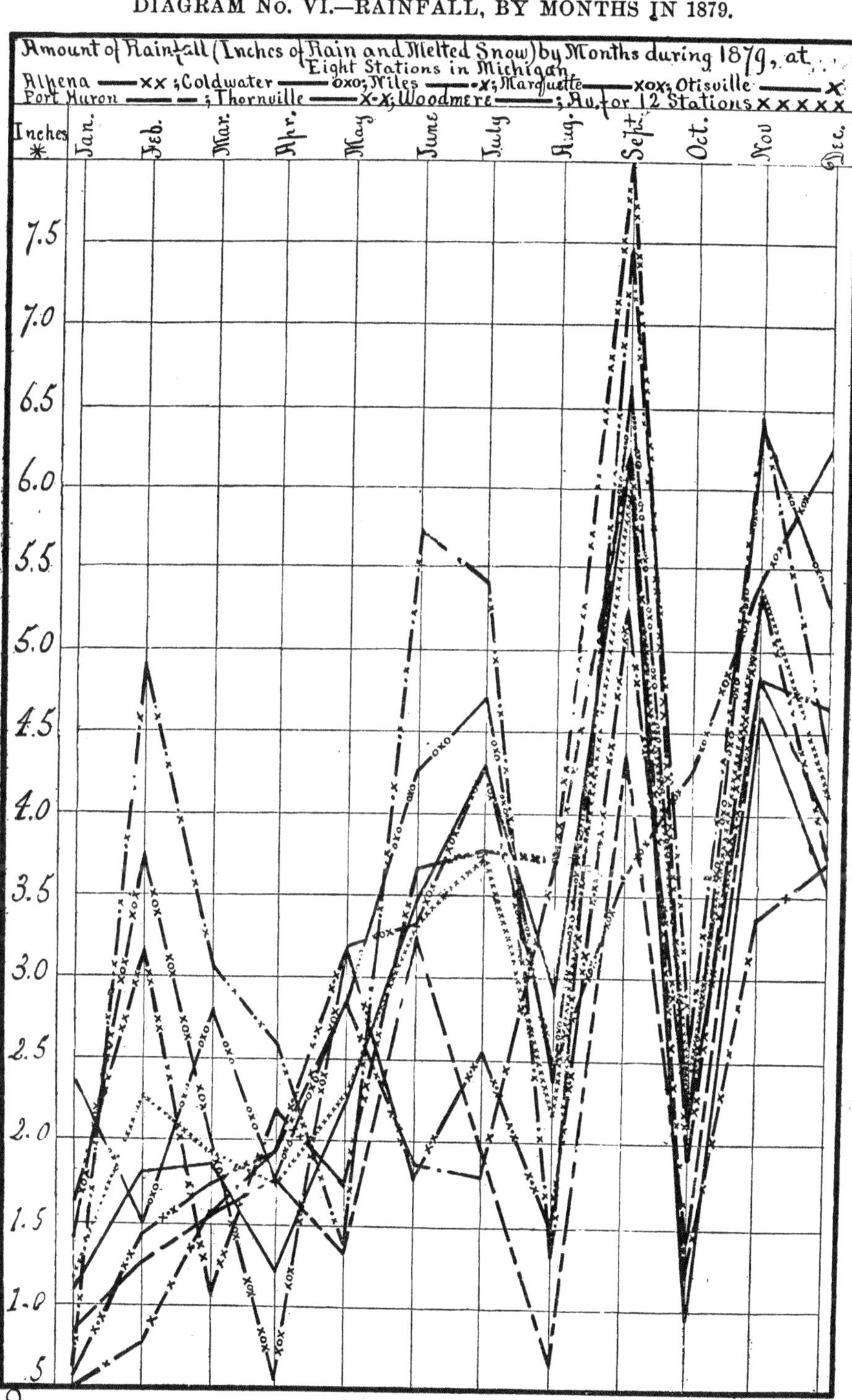

SCALE, ONE INCH OF RAINFALL TO .88 OF AN INCH, VERTICALLY.

Designed by Henry B. Baker. Drawn by H. B. Turner and Jno. K. Allen.

EXHIBIT 20.—*Comparison of the Average Amount of Atmospheric Ozone, as Indicated by the Degree of Coloration* of Schönbein's Test-paper, during the Year and during each Month of the Year 1879, with the Average for the Seven Years 1872–8, and with the Average for 1878; also Statement of the Average for each of the Seven Years 1872–78,—Test-paper exposed from 7 A. M. to 2 P. M., for the Day Observation, and from 9 P. M. to 7. A. M., for the Night Observation,—Observations made by Prof. R. C. Kedzie, at the State Agricultural College,† near Lansing, Mich.*

MONTHS.	OZONE,—DEGREES OF COLORATION.*																			
	DAY OBSERVATION,—7 A. M. TO 2 P. M.*										NIGHT OBSERVATION,—9 P. M. TO 7 A. M.*									
	1872.	1873.	1874.	1875.	1876.	1877.	1878.	For 7 Years, 1872–8.	1879.†	In 1879 More (+), or Less (-), than Av. for 7 Yrs., 1872–8.	In 1879 More (+), or Less (-), than in 1878.	1872.	1873.	1874.	1875.	1876.	1877.	1878.	For 7 Years, 1872–8.	1879.†	In 1879 More (+), or Less (-), than Av. for 7 Yrs., 1872–8.	In 1879 More (+), or Less (-), than in 1878.
Annual Av.	2.82	4.02	4.10	4.53	3.68	3.66	3.50	3.76	3.24	-.52	-.26	4.24	4.72	3.99	5.04	4.11	3.33	3.42	4.11	3.36	-.75	-.06
January	4.50	5.06	5.10	5.52	5.32	4.87	3.74	4.87	4.00	-.87	+.26	5.44	6.32	6.39	5.93	6.90	5.03	4.48	5.73	4.48	-1.30	0
February	4.00	4.21	5.53	5.78	5.76	4.54	5.00	4.97	4.54	-.43	-.46	6.00	5.93	6.10	6.57	7.27	4.14	5.33	5.91	5.01	-.90	-.32
March	3.00	4.09	5.00	5.70	6.06	5.29	4.68	4.83	3.93	-.90	-.75	5.71	5.80	6.00	6.19	7.00	6.06	4.94	5.95	4.30	-1.66	-.64
April	2.63	3.76	5.70	3.73	3.70	3.70	4.37	3.94	3.00	-.94	-1.37	5.30	4.27	6.60	3.96	5.53	4.27	4.53	4.92	4.00	-.92	-.53
May	2.14	4.58	5.80	3.19	3.10	3.29	3.97	3.72	2.79	-.93	-1.18	3.84	5.59	4.90	4.03	4.13	2.94	4.10	4.08	2.68	-1.40	-1.42
June	1.91	3.26	3.50	3.96	1.87	3.17	3.17	2.98	2.43	-.55	-.74	3.23	2.66	2.34	3.66	2.27	2.90	3.00	2.87	2.03	-.84	-.97
July	.88	3.59	1.80	3.52	2.16	2.71	2.10	2.39	2.90	+.51	+.80	2.28	3.06	1.50	3.22	1.58	1.77	1.74	2.16	2.13	-.03	+.39
August	.98	3.55	2.80	3.50	1.97	3.42	2.32	2.65	2.58	-.07	+.26	.99	2.65	1.10	3.70	1.26	1.03	1.52	1.75	1.76	+.01	+.24
September	1.40	2.30	2.90	4.30	2.60	2.93	2.23	2.67	2.87	+.20	+.64	1.86	4.27	1.20	3.68	1.73	1.53	1.60	2.27	2.53	+.26	+.93
October	1.93	4.52	3.32	5.30	3.45	3.20	2.71	3.49	3.32	-.17	+.61	3.18	4.69	3.01	6.20	3.48	3.19	1.97	3.67	2.81	-.86	+.84
November†	4.60	4.30	3.80	4.90	3.57	3.60	3.17	3.99	3.53	-.46	+.36	6.17	5.47	4.50	6.60	3.33	3.73	3.14	4.71	4.63	-.08	+1.49
December†	5.81	4.97	4.00	5.00	4.57	3.17	4.55	4.58	2.97	-1.61	-1.58	6.84	5.89	4.20	6.70	4.81	3.37	4.74	5.22	3.87	-1.35	-.87

* According to a scale of 10 degrees of coloration of Schönbein's test-paper. Maximum=10. The tinted scale is printed on page 142 of the Report of the State Board of Health for 1875.

† For November and December, 1879, the observations were made by Harry B. Turner, at the office of the State Board of Health, Lansing.

TABLE VIII.—*Relative Amount of Ozone in the Atmosphere, by Day during the Year and during each Month of the Year 1879, at 23 Stations in Michigan,—as Indicated by Averages of Observations made Daily by Exposing Test-paper prepared according to Schönbein's formula, from 7 A. M. to 2 P. M.—Recorded according to a scale of 10 Degrees of Coloration of the Test-paper (greatest coloration by Ozone equals 10), by Observers for the State Board of Health and for the U. S. Signal Service.**

STATIONS IN MICHIGAN.*	Geographical Divisions of the State.†	DEGREES OF COLORATION OF TEST-PAPER.—DAY OBSERVATIONS.													
		YEAR.		MONTHS, 1879.											
		1878.	1879.	Jan.	Feb.	Mar.	Apr.	May.	J'ne.	J'ly.	Aug.	Sept.	Oct.	Nov.	Dec.
AV. FOR 13 STATIONS ‡	....	[a]	2.85	3 62	3.58	3.49	3.12	2.85	2.24	2.28	2.30	2.48	2.44	2.87	2.89
Marquette *	U. P.†	.....	[b]	.....	.....	.....	.....	.....	.....	.....	.....	2.35	1.84	1.97	2.00
Menominee	U. P.	.....	[c]	.....	[l]3.43	3.30	2.97	2.77	2.07 [m]	1.55	1.47	0.89	2.65	3.31	.....
Petoskey	N.	2.22	2.56	2.00	2.39	2.10	2.23	1.65	1.90	2.61	3.23	3.23	2.94	3.20	3.29
Alpena	N. E.	.....	3.08	3.87	4.43	3.32	3.03	3 39	2.17	2.13	2.94	3.03	2.90	2.63	3.16
Grand Haven	W.	.....	[d]	.....	.....	.....	.....	.....	.....	.....	.....	2.87	3.03	2.77	3.03
Nirvana	W.	3.06	2.85	3.03	3.02	2.87	1.85	2.13	1.83	2.31	3.65	3.22	3.53	3.47	3.26
Port Huron	B. & E.	.....	[e]	.....	.....	.....	.....	.....	.....	.....	.....	3.07	2.74	.....	2.97
Thornville	B. & E.	2.78	1.77	4.23	2.64	2.39	1.87	1.26	1.50	1.52	0.42	1.03	0.77	1.60	2.03
Agr'l College	C.	3.50	[f]	4.00	4.54	3.93	3.00	2.79	2.43	2.90	2.58	2.87	3.32	.....	.....
Ionia	C.	.....	[g]	.....	.....	.....	.....	3.45	3.73	4.10	3.94	3.63	3.74	3.77	3.77
Lansing	C.	.....	3.15	[n]3.36	3.41	3.79	2.62	2.69	1.97	3.81	3.74	3.17	2.73	3.53	2.97
Otisville	C.	3.98	2.89	2.97	2.89	3.26	3.48	3.80	3.10	2.60	2.97	2.44	2.00	2.47	2.65
Benton Harbor	S. W.	.....	[h]	3.12	3.59	2.55	2.81	.....	.....	1.62	1.55	1.96	.....	2.00	2.10
Niles	S. W.	4.50	3.22	4.26	4.43	4.61	4.07	3.19	2.87	2.48	2.27	1.90	2.94	2.83	2.77
Battle Creek	S. C.	2,89	2.22	3.13	2.86	2.87	2.57	2.77	2.40	2.68	1.23	0.93	1.48	1.73	[m]1.93
Coldwater	S. C.	3.01	[i]	3.42	3.18	3.10	.....	.....	[o]3.29	3.03	2.12	2.66	[p]2.56	2.38 [m]	3.32
Kalamazoo	S. C.	2.67	2 49	2.84	2.71	2.65	2.60	2.65	2.17	1.94	1.94	2.50	2.39	2.83	2.65
Mendon	S. C.	3.08	2 67	4.31	4.75	[q]4.50	3.76 [m]	2.71	1.60	0.74	0.97	1.70	1.90	2.40	2.68
Tecumseh	S. C.	4.24	3.10	4 74	4.32	4.19	4.27	3.26	2.07	1.74	2.00	2.27	[q]2.48	3.27	2.55
Ypsilanti	S. C.	.....	[j]	.....	[p]3.54	3.27	2.27	2.42	[r]1.59	.....	[s]0.82	[r]1.20	1.68	[s]2.00	[n]2.58
Detroit	S. E.	.....	[k]	.....	.....	.....	0.01	0	.....	0	.....	0	0	0	0
Woodmere Cemetery.	S. E.	4.69	4.03	4.48	4.79	5.03	4.70	4.16	3.33	2.97	2.94	4 30	3.26	4.20	4.23
Washington	S. E.	.....	2.97	3.87	[r]3.84	3.73	3.50	3.42	2.22	2.06	1.65	2.52	2.37	3.12	3.34

* The names of observers, their places of observation, and the counties in which these places are situated, are stated in Exhibit 7, page 306.

† The full names of the divisions and the counties in each division are stated in Exhibit 1, page 227.

‡ An average for only the stations for which statements are given for every month of the year.

In the columns for months, the letters [l], [m], [n], etc., stand before or directly under the numbers from which they refer to the notes below.

[a] The average for 12 stations in 1878 is 3.39. [b] For 4 months, 2.04. [c] For 10 months, 2.44. [d] For 4 months, 2.93. [e] For 3 months, 2 93. [f] Including the observations at Lansing for November and December, the average of this line is 3.24. [g] For 8 months, 3.77. [h] For 9 months, 2.37. [i] For 10 months, 2.91. [j] For 10 months, 2.14. [k] For 7 months, 0.001. [l] For about 21 days. [m] For 29 days. [n] For 25 days. [o] For 17 days. [p] For 26 days. [q] For 30 days. [r] For 27 days. [s] For 28 days.

Graphic representations of 11 representative lines in this table are given in Diagram VII., opposite this page.

DIAGRAM NO. VII.—OZONE, DAY, BY MONTHS IN 1879.

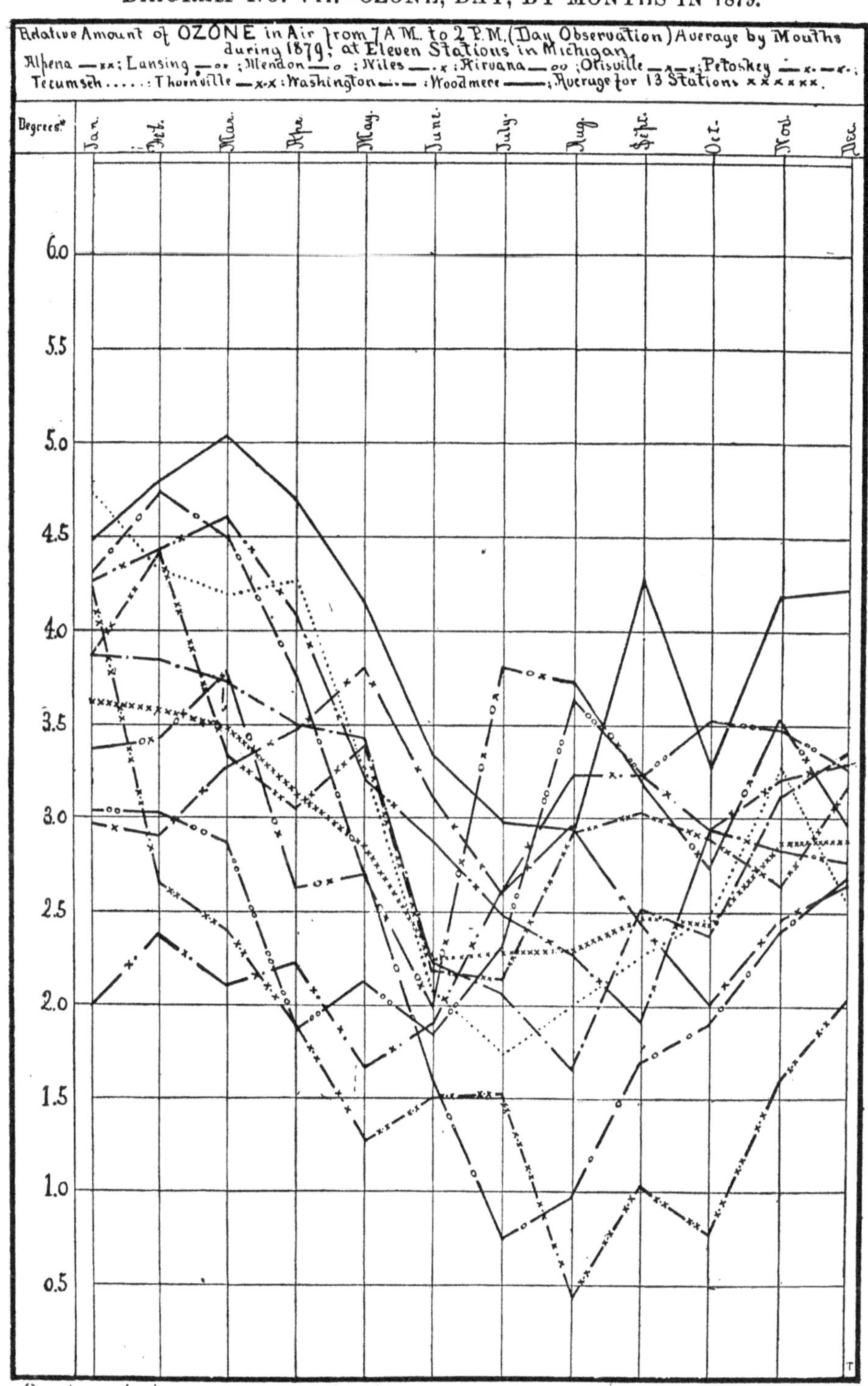

* One degree of coloration (on a scale of 10 degrees) to an inch, vertically.

Drawn by H. B. Turner and Jno. K. Allen.

Designed by Henry B. Baker.

DIAGRAM NO. VIII.—OZONE, NIGHT, BY MONTHS, IN 1879.

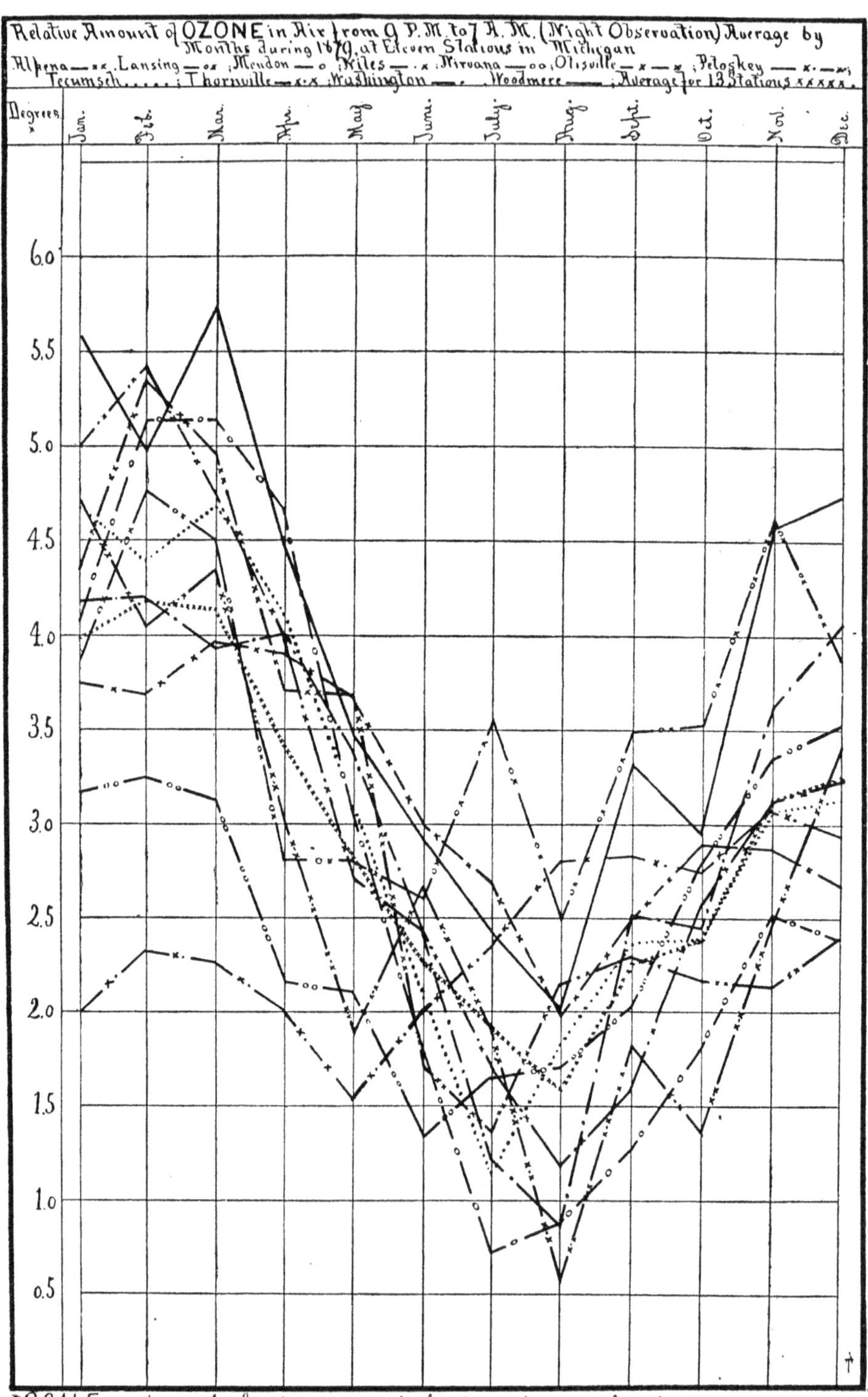

*SCALE, one degree of coloration (on a scale of 10 degrees) to an inch vertically.

Drawn by H. B. Turner and Jno. K. Allen.

Designed by Henry B. Baker.

TABLE IX.—*Relative Amount of Ozone in the Atmosphere at Night, during the Year and during each Month of the Year 1879, at 23 Stations in Michigan,—as indicated by Averages of Observations made Daily by exposing Test-paper, prepared according to Schönbien's formula, from 9 P. M. to 7 A. M.—Recorded, according to a scale of 10 Degrees of Coloration of the Test-paper (greatest coloration by Ozone equals 10), by Observers for the State Board of Health, and for the U. S. Signal Service.**

STATIONS IN MICHIGAN.*	GEOGRAPHICAL DIVISIONS OF THE STATE.†	DEGREES OF COLORATION OF TEST-PAPER.—NIGHT OBSERVATIONS.													
		YEAR.		MONTHS, 1879.											
		1878.	1879.	Jan.	Feb.	Mar.	Apr.	May.	June.	July.	Aug.	Sept.	Oct.	Nov	Dec.
Av. for 13 Stations‡	....	a	2.94	3.98	4.18	4.14	3.41	2.82	2.26	1.91	1.59	2.26	2.39	3.14	3.25
Marquette........*	U. P.†	.....	b	.....	.....	.....	....	.....	.....	.....	.....	2.12	1.98	2.17	2.10
Menominee........	U. P.	.....	c	....	[l] 4.65	4.00	3.77	2.81	2.02	1.37	1.05	0.86	3.07	3.90	.....
Petoskey..........	N.	2.11	2.40	2.00	2.32	2.26	2.00	[m]1.52	2.00	2.35	2.81	2.83	2.74	3.07	2.94
Alpena..............	N. E.	.....	3.02	4.35	5.36	4.97	3.70	3.68	1.70	1.35	2.16	2.30	2.16	2.13	2.39
Grand Haven......	W.	.....	d	.....	.....	.....	....	.....	.....	.....	.....	2.57	3.00	2.93	3.06
Nirvana............	W.	2.57	2.51	3.16	3.25	3.13	2.15	2.11	1.33	1.65	1.71	2.02	2.79	3.35	3.52
Port Huron........	B. & E.	.....	e	.....	.....	.....	....	.....	.....	.....	.....	3.03	2.71	.....	3.10
Thornville..........	B. & E.	3.09	2.68	4.71	4.04	4.35	3.00	1.87	2.67	1.90	0.55	1.83	1.35	2.47	3.42
Agrl. College, near Lansing..........	C.	3.42	f	4.48	5.01	4.30	4.00	2.68	2.03	2.13	1.76	2.53	2.81	.....	.....
Ionia...............	C.	.....	g	.....	.....	.....	....	3.20	3.57	3.74	4.16	3.93	3.71	3.73	4.06
Lansing............	C.	.....	3.58	[n]3.87	4.77	4.50	2.80	2.80	2.60	3.56	2.48	3.50	3.52	4.63	3.87
Otisville...........	C.	4.40	3.13	3.74	3.68	3.97	3.90	3.67	3.00	2.67	1.97	2.50	2.90	2.87	2.69
Benton Harbor....	S. W.	.....	h	3.23	4.07	3.27	3.04	.....	.....	1.96	1.54	2.33	.....	.....	2.97
Niles...............	S. W.	4.18	3.14	5.00	5.43	4.74	4.00	2.71	2.43	1.69	1.18	1.58	2.58	3.13	3.23
Battle Creek.......	S. C.	3.00	2.09	3.10	3.50	2.97	2.73	2.68	2.33	2.39	[*]0.42	0.67	0.68	1.43	[o] 2.13
Coldwater.........	S. C.	3.61	i	3.74	3.46	3.13	....	.....	[p] 3.59	3.79	4.08	2.96	[q]2.58	[o]2.69	2.87
Kalamazoo.........	S. C.	3.17	2.85	3.26	3.29	3.39	2.80	2.58	2.17	1.81	1.77	2.68	2.71	3.97	3.71
Mendon.............	S. C.	3.21	2.79	4.07	5.14	[o]5.14	4.67	3.06	1.79	0.71	[r]0.87	1.27	1.81	2.53	2.39
Tecumseh..........	S. C.	3.95	3.08	4.68	4.39	4.68	4.10	3.10	2.07	1.13	1.84	2.37	2.39	3.07	3.13
Ypsilanti...........	S. C.	.....	j	3.64	[n] 3.40	3.80	2.50	1.87	[m]1.67	.....	[s] 1.14	[o]1.76	1.71	[s] 1.96	[m]3.15
Detroit.............	S. E.	.....	k	0.16	0.92	.....	0.02	0	.....	0	.....	0	0	0	0
Woodmere Cemetery, near Detroit	S. E.	4.47	3.93	5.58	4.96	5.74	4.47	3.45	2.90	2.42	2.00	3.33	2.94	4.60	4.74
Washington.......	S. E.	.....	3.07	4.18	[m]4.20	3.92	4.00	3.39	2.42	1.21	0.85	2.52	2.45	3.63	4.05

* The names of observers, their places of observation, and the counties in which these places are situated, are stated in Exhibit 7, page 306.

† The full names of the divisions and the counties in each division, are stated in Exhibit 1, page 227.

‡ This line is an average for only the stations for which statements are given for every month of the year.

a The average for 12 stations in 1878 is 3.43. b For 4 months, 2.09. c For 10 months, 2.75. d For 4 months, 2.89. e For 3 months, 2.95. f Including the observations at Lansing for Nov. and Dec., the average of this line is 3.36. g For 8 months, 3.76. h For 8 months, 2.80. i For 10 months, 3.29. j For 11 months, 2.42. k For 9 months, 0.12. l For about 21 days. m For 27 days. n For the last 25 days. o For 29 days. p For 17 days. q For 26 days. r For 30 days. s For 28 days.

Graphic representations of 11 representative lines in this table are given in Diagram VIII., opposite this page.

VELOCITY AND DIRECTION OF THE WIND.

While the general course of the line representing the average velocity of the wind in 1879 (Diagram IX., page 351) is similar to that of the line for 1878 (Diagram VIII., page 375 of the Report for 1879), the lowest point for 1879 is the same as the highest point for 1878, representing a velocity of 6.8 miles per hour. This great difference may be in part explained by the fact that in March, 1879, the anemometer was moved from the roof of the old "State-Offices" building to the roof of the new capitol building, a much higher position.

The least average velocity in 1879 occurred in June, whereas in 1878 it was in July and August. In October, 1879, the wind was, as compared with other months of the year, lower than in October, 1878, though the average velocity actually recorded was greater in October, 1879, than in October, 1878, by four miles per hour. This relatively low wind in October, 1879, should be considered in connection with the high temperature of that month. By comparing Table XI., page 354, with Table X., page 380 of the Report for 1879, or Diagram XI., page 355, with Diagram X., page 381 of the Report for 1879, it would seem that winds from the north, northwest, west, and southwest, were slightly less prevalent, and winds from the other four main points of compass slightly more prevalent in October 1879 than in October 1878.

By the tables and diagrams relating to the direction of the wind, pages 354–60, it will be noticed that some observers report a much larger proportion of calms than do others. This is probably in part due to a difference in judgment of observers as to what shall be regarded as a calm. Observers may do well to compare the summaries of their own reports with those of others.

The most notable occurrence during the month was the heavy wind storm of the 19th and 20th. From 4:15 of the 19th to 4:15 of the 20th, there were 640 miles of wind passed over the station.—*J. A. Barwick, Serg't Sig. Corps, U. S. A., Alpena, on November register.*

DIAGRAM NO. IX.—VELOCITY OF THE WIND, BY MONTHS IN 1879.

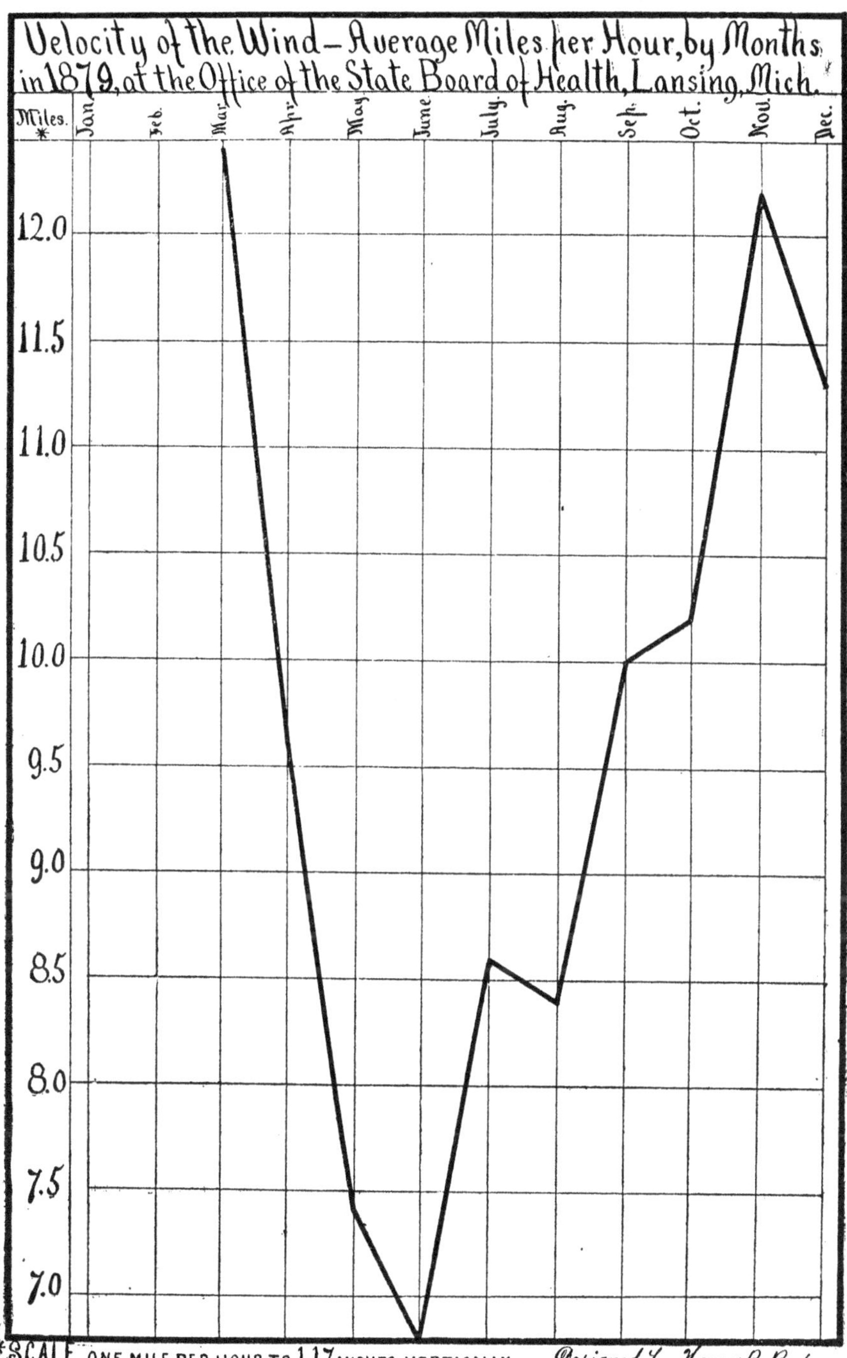

*SCALE, ONE MILE PER HOUR TO 1.17 INCHES VERTICALLY. Designed by Henry B. Baker.

Drawn by H. B. Turner and Jno. K. Allen.

TABLE X.—*Average Velocity of the Wind, in Miles per Hour, during each Hour of the Day, by Months, for Ten Months of the Year 1879.—Compiled from Registers of the Robinson's Self-Registering Anemometer in the Office of the State Board of Health, State Capitol, Lansing, Michigan.*

MONTHS,—1879.	HOURS, AND MILES PER HOUR.																								
	AVERAGE.	A. M.					P. M.												A. M.						
		7–8	8–9	9–10	10–11	11–12	12–1	1–2	2–3	3–4	4–5	5–6	6–7	7–8	8–9	9–10	10–11	11–12	12–1	1–2	2–3	3–4	4–5	5–6	6–7
AV., 10 MONTHS	9.7	8.7	9.5	10.4	11.3	11.7	12.5	12.6	12.8	12.6	12.0	10.6	9.3	8.8	8.5	8.4	8.3	7.9	8.2	7.8	7.9	7.9	8.0	7.8	8.5
March [a]	12.4	11.2	12.5	14.1	14.8	14.2	15.6	16.0	14.9	15.1	14.3	13.0	11.8	11.2	10.3	9.3	10.0	10.6	11.5	10.7	11.2	11.3	10.9	11.3	11.5
April	9.6	9.1	10.2	10.4	11.0	12.1	12.2	12.0	13.5	13.5	12.3	11.2	10.0	8.3	7.9	7.7	7.1	6.9	7.3	7.5	7.7	7.8	8.3	8.0	7.7
May	7.4	6.4	6.9	7.6	8.9	9.0	9.8	9.4	10.4	10.6	11.1	9.0	6.4	6.3	5.8	6.0	6.1	6.1	6.2	5.3	6.3	6.2	5.9	5.1	5.7
June	6.8	5.8	6.5	7.7	8.0	8.8	8.6	8.4	8.7	8.6	7.7	7.4	6.1	6.1	6.7	6.9	6.3	6.0	6.4	5.8	5.6	5.2	5.6	4.9	5.3
July	8.6	7.8	9.0	9.3	9.9	10.0	11.1	11.7	12.2	12.2	13.4	12.2	9.5	7.5	6.8	7.0	6.8	5.9	5.7	5.8	6.1	6.3	6.4	6.2	6.9
August	8.4	6.6	7.9	9.3	10.4	11.8	12.5	12.7	12.9	12.5	12.0	10.9	8.2	7.4	7.0	7.2	6.4	6.1	6.6	5.9	5.5	5.6	5.6	5.5	5.7
September [b]	10.0	7.5	9.1	10.9	12.1	12.0	12.3	12.4	12.8	12.7	11.9	9.6	9.0	8.9	9.3	9.3	9.8	9.1	9.6	9.2	8.2	8.1	8.5	8.1	8.9
October [c]	10.2	9.6	9.9	10.3	11.0	12.2	13.6	13.3	13.4	12.8	11.4	9.4	9.4	9.4	9.7	9.5	9.2	8.8	8.6	8.6	8.8	8.2	8.6	8.5	9.4
November	12.2	12.0	11.9	12.8	13.8	13.5	14.6	15.9	15.5	14.7	13.6	11.9	11.8	12.0	11.6	10.9	10.4	9.5	10.3	9.9	10.2	10.8	10.6	11.2	12.8
December	11.3	11.4	11.2	11.7	13.2	13.2	14.6	13.9	13.8	13.0	12.1	11.2	11.2	10.5	10.2	10.2	10.5	10.0	9.7	9.7	9.4	9.1	9.8	9.4	11.2

[a] For only about 27 days. [b] For only about 28 days. [c] For only about 30 days.

The statements in the first figure-column in Table X., of the average velocity of the wind in miles per hour, are graphically represented for ten months of the year 1879, in Diagram IX., page 351. The remaining columns of Table X. are graphically represented in Diagram X., page 353.

DIAGRAM No. X.—VELOCITY OF THE WIND AT EACH HOUR, 1879.

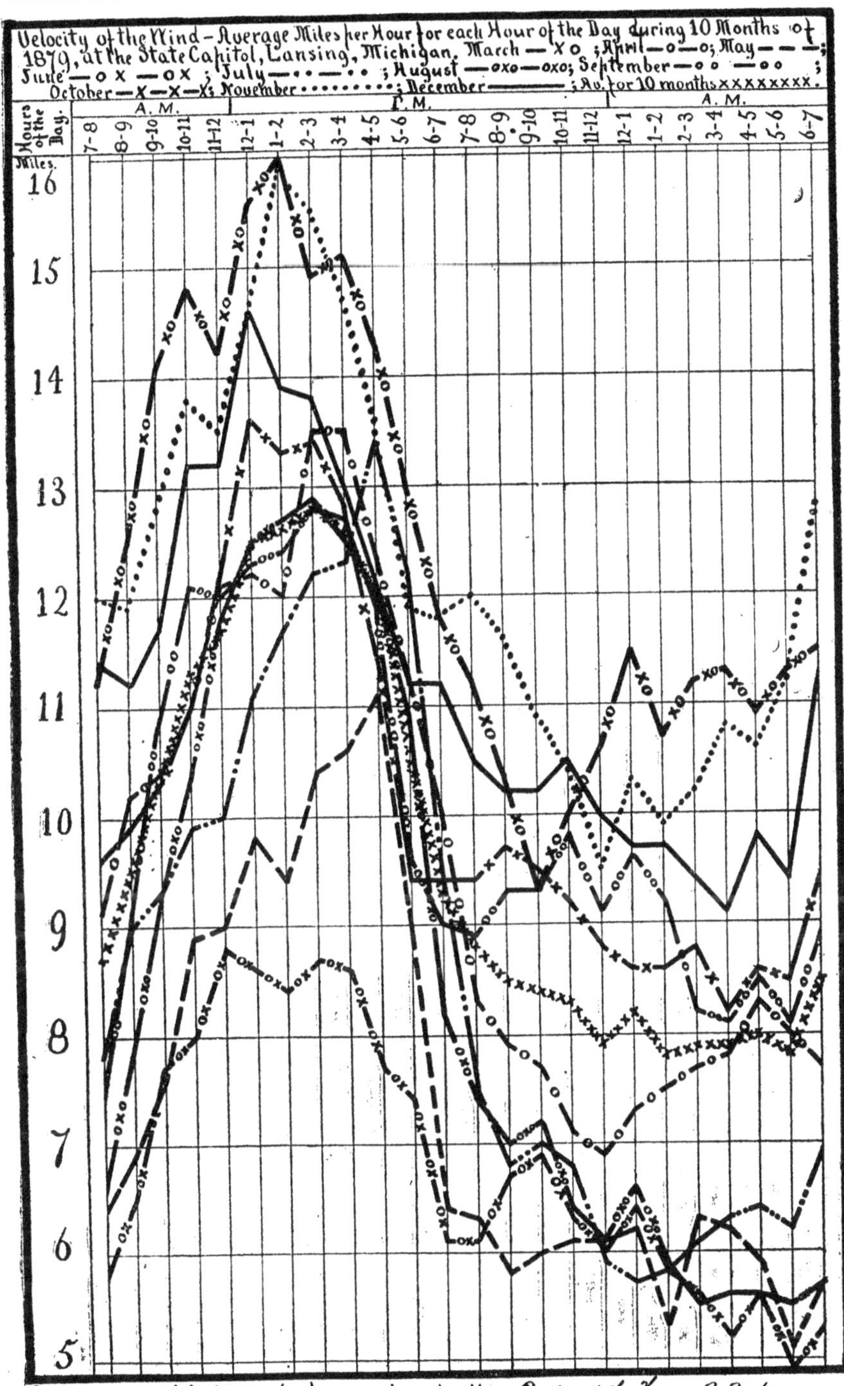

SCALE, one mile per hour to .6 of an inch, vertically. *Designed by Henry B. Baker.*
Drawn by H. B. Turner and Jno. K. Allen.

TABLE XI.—*Number of Observations per Month (at 7 A. M., 2 P. M., and 9 P. M., daily), at which the Wind was Blowing from each of the Eight Principal Points of Compass, during the Year and during each Month of the Year 1879,—Average for 16 Stations in Michigan.**

POINTS OF COMPASS.	AVERAGE NUMBER OF OBSERVATIONS PER MONTH, 1879.												
	Year.	Jan.	Feb.	Mar.	Apr.	May.	June.	July.	Aug.	Sept.	Oct.	Nov.	Dec.
All Observations.	91	93	84	93	90	93	90	93	93	90	93	90	92
Calm	6	4	4	4	6	6	6	7	8	6	7	4	4
North	6	3	5	6	10	7	8	6	6	5	4	4	4
North-east	7	3	6	6	11	11	7	8	10	4	4	5	8
East	5	3	5	7	5	6	4	3	5	5	3	3	7
South-east	9	6	11	12	9	13	6	7	5	7	10	9	11
South	10	11	7	7	6	10	11	9	9	12	16	15	8
South-west	18	29	11	13	12	14	11	25	22	18	22	19	20
West	16	18	17	19	9	12	12	18	15	17	16	17	18
North-west	16	15	19	19	22	14	18	10	13	17	11	15	13

Graphic representations of statements in Tables XI. and XII. are given in Diagrams XI. and XII., opposite this page.

TABLE XII.—*Average Number of Observations per Month, for the Year 1879, at which the Wind was Blowing from each of the Eight Principal Points of Compass, at each of 16 Stations in Michigan; also the Average for All said Stations.**

STATIONS IN MICHIGAN.*	Divisions of the State.†	AVERAGE NUMBER OF OBSERVATIONS PER MONTH, IN 1879.									
		All Obs.	Calms.	N.	N. E.	E.	S. E.	S.	S. W.	W.	N. W.
Average, 16 Stations	----------	91	6	6	7	5	9	10	18	16	16
Marquette*	U. P.†	91	6	6	5	4	8	6	15	14	26
Petoskey	N.	91	4	1	2	1	9	5	29	19	23
Alpena	N. E.	91	5	6	5	4	17	8	10	16	20
Grand Haven	W.	91	4	10	8	10	6	11	15	14	14
Nirvana	W.	91	4	4	14	9	9	2	32	6	11
Port Huron	B. & E.	91	2	13	12	4	6	21	13	12	9
Thornville	B. & E.	91	15	2	9	2	12	7	20	12	12
Otisville	C.	91	1	3	6	5	17	5	20	12	21
Niles	S. W.	91	0	3	2	2	6	5	22	32	20
Battle Creek	S. C.	91	2	3	8	4	10	14	8	20	14
Kalamazoo	S. C.	91	1	4	6	2	12	14	28	8	17
Mendon	S. C.	89	4	6	7	5	8	21	14	16	11
Tecumseh	S. C.	91	10	6	4	5	5	8	14	21	17
Detroit	S. E.	91	1	8	10	8	4	8	26	16	10
Woodmere Cemetery	S. E.	91	3	8	8	6	3	17	14	22	12
Washington	S. E.	91	29	8	6	3	5	7	9	13	11

* The names of observers, their places of observation, and the counties, and divisions of the State, in which these places are situated, are stated in Exhibit 7, page 306.

† The full names of the divisions, and the counties in each division, are stated in Exhibit 1, page 227.

DIAGRAM No. XI.—DIRECTION OF THE WIND IN MICHIGAN, BY MONTHS IN 1879.

NORTH.

Explanations of the construction and manner of reading Diagrams XI., XII., and XIII., are given on pages 307–8. The top of each diagram is North; and the scale is such that each one-hundredth of an inch on the line represents one observation of the wind from the direction indicated by the line. Calms are represented by circles. Comments on the subject of the direction of the wind, and on calms, are printed on page 350.

DIAGRAM No. XII.—DIRECTION OF THE WIND AT STATIONS IN MICHIGAN, 1879.

NORTH.

DIAGRAM XIII.—DIRECTION OF WIND, AT STATIONS, MONTHS IN 1879.

NORTH.

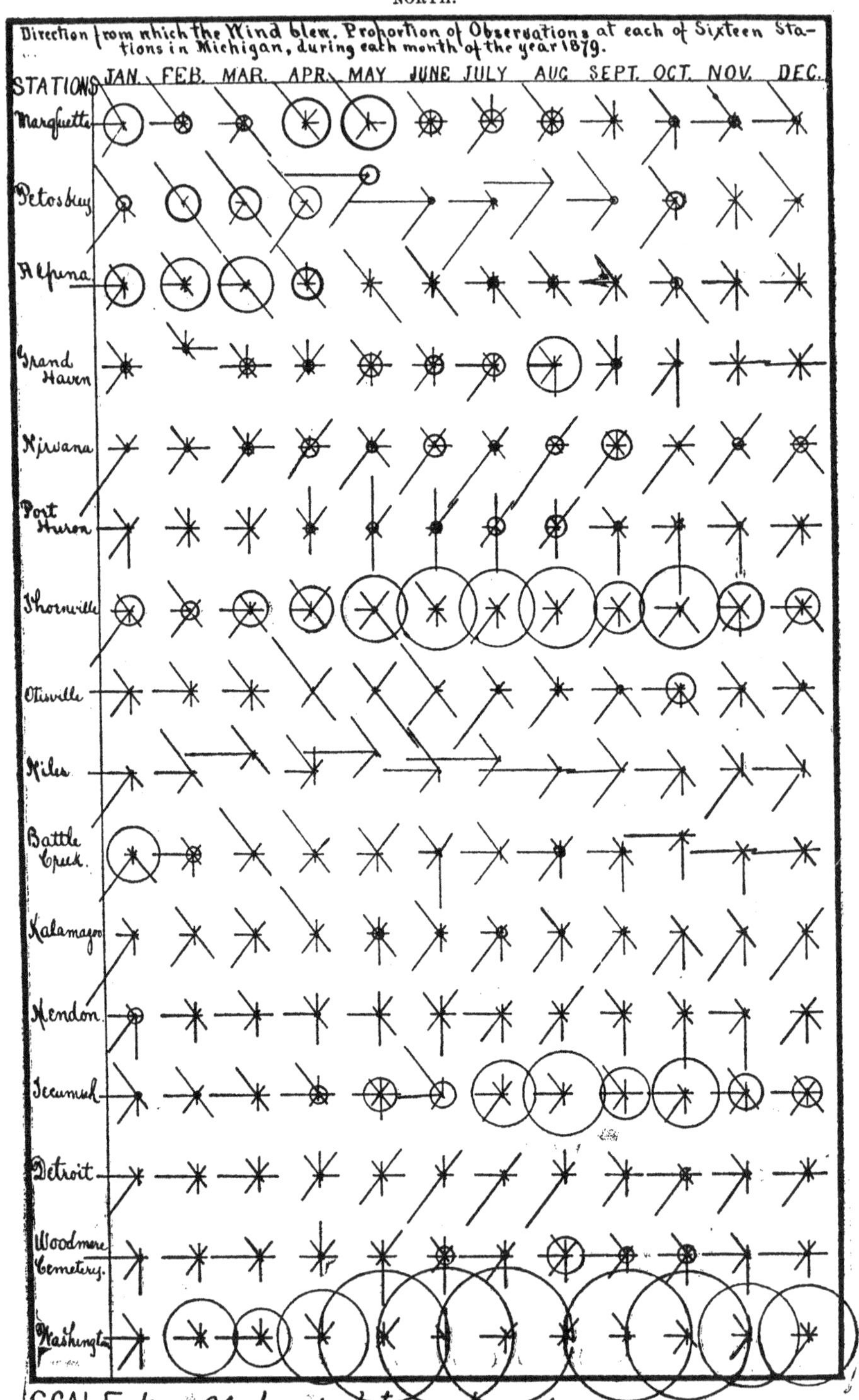

TABLE XIII.—*Number of Observations for each Month of the Year 1879, at which the Wind was Blowing from each of the Eight Principal Points of Compass, at each of 23 Stations* in Michigan; also the Average for the 16 of said Stations from which nearly Complete Observations were received for the Year. (Observations made at 7 A. M., 2 P. M., and 9 P. M., Daily.)*

STATIONS IN MICHIGAN.*	DIVISIONS OF THE STATE.†	JANUARY.										FEBRUARY.										MARCH.									
		Total.	Calm.	N.	N. E.	E.	S.E.	S.	S. W.	W.	N.W.	Total.	Calm.	N.	N. E.	E.	S.E.	S.	S. W.	W.	N.W.	Total.	Calm.	N.	N. E.	E.	S.E.	S.	S. W.	W.	N.W.
Average for 16 Stations....‡	------	93	4	3	3	3	6	11	29	18	15	84	4	5	6	5	11	7	11	17	19	93	4	6	6	7	12	7	13	19	19
Marquette ....*	U.P.†	93	11	2	2	0	4	3	22	19	30	84	5	7	4	3	6	0	5	20	34	93	4	4	4	0	24	3	13	15	26
Menominee	U. P.	93	0	14	5	9	6	12	6	33	8	84	0	19	6	3	9	3	3	18	23	93	0	8	11	7	13	10	4	18	22
Petoskey	N.	93	4	1	1	1	10	11	35	1	29	84	10	0	5	0	35	3	1	1	29	93	9	0	0	1	24	2	15	5	37
Alpena..........	N. E.	93	11	5	2	0	3	8	17	33	14	84	14	4	6	4	9	5	5	14	23	93	15	1	3	3	24	3	6	17	21
Grand Haven...	W.	93	3	9	6	10	7	7	19	14	18	84	2	11	8	19	9	4	3	9	19	93	4	12	6	14	8	7	9	17	16
Nirvana.........	W.	93	1	3	10	7	2	2	47	11	10	84	2	6	8	13	9	0	19	12	15	93	3	8	13	15	4	5	25	6	14
Port Huron......	B.& E.	93	0	3	8	3	3	22	21	19	14	84	0	11	6	5	9	11	13	14	15	93	0	12	14	3	8	15	13	15	13
Thornville......	B.& E.	93	8	0	4	1	11	7	39	11	12	84	5	1	9	2	12	3	19	12	21	93	10	4	4	7	16	3	21	18	10
Agrl. College...	C.	89	23	0	2	2	5	2	42	7	6	84	19	0	7	0	10	3	9	7	29	93	20	0	5	0	9	5	29	6	19
Ionia..........	C.																														
Lansing.........	C.	73	24	0	4	0	1	6	25	8	5	84	30	0	3	0	7	6	9	5	24	87	21	0	2	4	7	16	15	15	7
Otisville	C.	93	0	2	3	7	9	13	18	25	16	83	0	6	5	4	8	8	8	23	21	93	0	6	5	10	11	11	11	17	22
Benton Harbor.	S.W.	91	15	6	1	2	8	24	9	20	6	84	7	1	2	7	12	13	7	20	12	87	15	10	2	9	3	11	17	10	10
Niles	S.W.	93	0	0	1	3	9	10	39	26	5	84	0	0	1	3	10	1	15	23	31	93	0	0	2	2	15	3	11	40	20
Battle Creek ...	S. C.	93	15	4	3	2	20	5	33	6	5	84	4	2	7	5	7	12	11	24	12	93	1	2	7	6	15	5	13	12	32
Coldwater	S. C.	93	0	1	0	0	0	7	67	16	2	84	0	5	4	4	2	13	28	26	2	48	0	1	3	7	0	11	8	14	4
Kalamazoo......	S. C.	93	0	2	3	2	3	12	49	11	11	84	0	4	5	0	17	6	23	10	19	93	0	3	9	3	12	11	24	13	18
Mendon.........	S. C.	87	4	1	1	4	2	25	26	16	8	83	0	7	5	3	10	16	10	21	11	87	0	7	3	9	3	12	11	25	17
Tecumseh	S. C.	93	2	2	2	1	8	12	20	23	23	84	2	2	6	1	16	5	13	20	19	93	1	6	6	6	10	9	6	29	20
Ypsilanti........	S. C.	89	0	2	3	0	2	6	45	27	4	84	0	1	9	2	11	3	20	15	20	93	0	1	6	6	14	8	12	27	19
Detroit..........	S. E.	93	0	5	5	3	3	5	39	25	8	84	0	7	7	4	7	10	12	26	11	93	0	11	9	11	4	9	13	24	12
Woodmere Cem.	S. E.	93	0	5	2	4	1	21	15	33	12	84	1	8	7	5	5	14	13	22	9	93	0	4	10	8	3	14	11	29	14
Washington.....	S. E.	93	0	5	2	5	2	17	23	22	17	84	21	8	5	3	7	6	6	16	12	93	16	11	6	6	10	4	5	27	8

* The names of observers, their places of observation, and the counties in which these places are situated, are stated in Exhibit 7, page 306.
† The full names of the divisions and the counties in each division, are stated in Exhibit 1, page 227.
‡ This line is an average for only the 16 stations from which statements nearly complete were received for every month of the year; it does not include the lines for Menominee, Agricultural College, Lansing Ionia, Benton Harbor, Coldwater, or Ypsilanti.

Graphic representations of statements for 16 lines in this table are given in Diagram XIII., page 356, which is explained on pages 307–8.

TABLE XIII.—CONTINUED.—*Direction of Wind, Months in 1879.—Observations at which the Wind was Blowing from Directions Named.*

APRIL.

STATIONS IN MICHIGAN.*	DIVISIONS OF THE STATE.†	Total.	Calm.	N.	N.E.	E.	S.E.	S.	S.W.	W.	N.W.
Average for 16 Stations....‡	......	90	6	10	11	5	9	6	12	9	22
Marquette.	U. P.†	90	13	7	10	7	12	4	5	3	29
Menominee	U. P.	90	0	10	41	15	10	3	2	3	6
Petoskey........	N.	90	9	0	2	0	7	0	35	4	33
Alpena..........	N. E.	90	8	12	10	5	17	4	1	6	27
Grand Haven...	W.	90	3	15	15	10	6	5	12	9	15
Nirvana.........	W.	90	5	6	20	11	5	1	22	7	13
Port Huron.....	B.&E.	90	3	23	15	4	4	15	7	5	14
Thornville......	B.&E.	90	12	4	16	0	12	4	9	12	21
Agrl. College...	C.	89	17	1	14	0	6	3	16	4	28
Ionia............	C.	......	......	..	.....	..	....	..	.....	...	.....
Lansing.........	C.	88	33	6	3	1	3	3	10	8	21
Otisville	C.	90	0	1	16	0	21	0	12	1	39
Benton Harbor.	S. W.	81	24	16	8	6	5	5	5	6	6
Niles	S. W.	90	0	14	8	2	1	8	13	18	26
Battle Creek....	S. C.	90	0	2	12	5	14	5	14	7	31
Coldwater.......	S. C.	89	0	1	1	5	30	7	10	11	24
Kalamazoo......	S. C.	90	0	8	6	1	13	11	14	4	33
Mendon.........	S. C.	86	1	12	11	1	9	17	2	20	13
Tecumseh	S. C.	90	5	10	2	9	7	3	9	23	22
Ypsilanti........	S. C.	90	0	9	13	4	8	8	17	17	14
Detroit..........	S. E.	90	1	14	16	11	3	7	18	5	15
Woodmere Cem.	S. E.	90	2	19	12	9	2	11	11	12	11
Washington	S. E.	90	26	12	11	4	5	2	4	10	16

MAY.

STATIONS IN MICHIGAN.*	DIVISIONS OF THE STATE.†	Total.	Calm.	N.	N.E.	E.	S.E.	S.	S.W.	W.	N.W.
Average for 16 Stations....‡	......	93	6	7	11	6	13	10	14	12	14
Marquette.	U. P.†	93	15	6	1	10	7	5	12	2	35
Menominee	U. P.	93	0	9	30	13	22	19	0	0	0
Petoskey........	N.	93	5	0	0	0	0	1	37	48	2
Alpena..........	N. E.	93	1	8	4	7	28	8	7	6	24
Grand Haven...	W.	93	6	7	10	12	7	12	13	8	18
Nirvana.........	W.	93	3	5	12	11	15	2	30	2	13
Port Huron.....	B.&E.	93	3	29	16	1	7	24	8	4	1
Thornville......	B.&E.	93	18	1	12	1	21	3	14	11	12
Agrl. College...	C.	93	37	2	10	3	8	6	14	5	8
Ionia............	C.	90	0	2	28	5	8	5	16	8	18
Lansing.........	C.	83	23	5	2	0	2	6	16	23	6
Otisville	C.	93	0	1	19	3	40	0	9	1	20
Benton Harbor.	S. W.	......	......	..	.....	..	....	..	.....	...	.....
Niles	S. W.	93	0	0	0	0	4	2	20	42	25
Battle Creek....	S. C.	93	0	0	20	3	13	11	16	12	18
Coldwater.......	S. C.	93	0	2	21	6	10	8	29	8	9
Kalamazoo......	S. C.	93	3	9	7	6	16	18	14	7	13
Mendon.........	S. C.	91	0	13	13	6	15	17	2	19	6
Tecumseh	S. C.	93	9	9	8	11	8	14	8	16	10
Ypsilanti........	S. C.	93	0	5	11	14	18	12	13	8	12
Detroit..........	S. E.	93	0	8	19	13	4	12	21	6	10
Woodmere Cem.	S. E.	93	0	9	18	8	6	27	6	9	10
Washington	S. E.	93	37	8	14	2	9	5	7	4	7

JUNE.

STATIONS IN MICHIGAN.*	DIVISIONS OF THE STATE.†	Total.	Calm.	N.	N. E.	E.	S. E.	S.	S.W.	W.	N.W.
Average for 16 Stations....‡	......	90	6	8	7	4	6	11	11	12	18
Marquette.	U. P.†	90	6	7	5	10	9	9	12	10	22
Menominee	U. P.	90	0	0	23	12	22	28	5	0	0
Petoskey........	N.	90	2	1	2	0	0	0	28	47	10
Alpena..........	N. E.	90	2	7	6	4	23	8	7	7	26
Grand Haven...	W.	90	5	10	8	5	2	13	19	13	15
Nirvana.........	W.	90	6	3	11	10	7	1	35	4	13
Port Huron.....	B.&E.	90	3	22	20	1	4	24	7	6	3
Thornville......	B.&E.	90	23	3	9	4	9	10	10	5	17
Agrl. College...	C.	90	25	11	6	3	5	7	22	5	6
Ionia............	C.	90	0	6	27	0	6	5	20	6	20
Lansing.........	C.	89	2	9	15	0	7	15	21	12	8
Otisville	C.	90	0	2	7	3	19	0	25	0	34
Benton Harbor.	S. W.	......	......	..	.....	..	....	..	.....	...	.....
Niles	S. W.	90	0	1	0	0	2	7	17	33	30
Battle Creek....	S. C.	90	0	2	13	0	1	31	12	12	19
Coldwater.......	S. C.	90	0	4	11	7	13	11	27	8	9
Kalamazoo......	S. C.	87	2	5	5	4	6	11	27	4	23
Mendon.........	S. C.	89	0	10	2	8	9	24	13	7	16
Tecumseh	S. C.	90	7	9	0	0	2	1	10	26	35
Ypsilanti........	S. C.	86	0	17	12	2	2	18	14	10	11
Detroit..........	S. E.	90	1	11	17	6	2	7	35	4	7
Woodmere Cem.	S. E.	90	5	11	7	6	4	21	12	11	13
Washington	S. E.	90	39	13	4	1	2	10	2	8	8

* The names of observers, their places of observation, and the counties in which these places are situated, are stated in Exhibit 7, page 306.
† The full names of the divisions and the counties in each division are stated in Exhibit 1, page 227.
‡ This line is an average for only the 16 stations from which statements nearly complete were received for every month of the year; it does not include the lines for Menominee, Agricultural College, Lansing, Ionia, Benton Harbor, Coldwater, or Ypsilanti.

Graphic representations of statements for 16 lines in this table are given in Diagram XIII., page 356, which is explained on pages 307–8.

TABLE XIII.—CONTINUED.—*Direction of Wind, Months in 1879.—Observations at which the Wind was Blowing from Directions Named.*

STATIONS IN MICHIGAN.*	DIVISIONS OF THE STATE.†	JULY.										AUGUST.										SEPTEMBER.									
		Total.	Calm.	N.	N. E.	E.	S.E.	S	S.W.	W.	N.W.	Total.	Calm.	N.	N. E.	E.	S.E.	S.	S.W.	W.	N.W.	Total.	Calm.	N.	N. E.	E.	S.E.	S.	S.W.	W.	N.W.
Average for 16 Stations....‡	------	93	7	6	8	3	7	9	25	18	10	93	8	6	10	5	5	9	22	15	13	90	6	5	4	5	7	12	18	17	17
Marquette......	U. P.†	93	6	15	11	6	4	2	23	8	18	93	6	8	10	6	5	8	14	9	27	90	1	11	4	5	8	11	12	20	18
Menominee......	U. P.	93	0	0	24	10	14	22	17	6	0	93	6	0	19	5	30	8	20	3	2	90	4	2	19	10	24	17	14	0	0
Petoskey	N.	93	1	0	0	2	3	5	48	26	8	93	0	0	0	0	0	0	41	41	11	90	2	0	1	0	1	1	22	28	35
Alpena..........	N. E.	93	3	6	5	7	25	12	6	17	12	93	3	5	5	10	17	9	7	15	22	90	0	4	6	2	11	11	16	21	19
Grand Haven...	W.	93	6	6	6	3	5	10	26	22	9	93	15	5	7	4	2	19	13	15	13	90	3	16	5	3	2	12	19	12	18
Nirvana..........	W.	93	3	1	13	5	10	5	43	4	9	93	5	0	21	8	6	0	44	4	5	90	8	7	9	5	7	7	31	6	10
Port Huron.....	B.&E.	93	5	19	17	0	3	25	15	9	0	93	6	15	21	3	3	17	16	6	6	90	3	3	7	6	7	27	6	17	14
Thornville......	B.&E.	93	21	1	10	2	8	11	25	7	8	93	22	4	14	1	4	10	17	9	12	90	14	0	10	0	10	10	29	9	8
Agrl. College...	C.	92	34	5	2	1	2	4	21	17	6	92	34	11	7	2	1	4	27	3	3	90	31	2	7	0	7	7	14	7	15
Ionia	C.	93	0	4	19	4	8	3	16	21	18	93	0	3	25	0	4	2	22	16	21	90	1	7	7	6	9	6	18	15	21
Lansing..........	C.	92	0	0	14	5	6	14	29	15	9	93	1	5	17	1	4	12	33	8	12	90	4	4	5	3	10	21	19	13	11
Otisville	C.	92	1	2	5	7	12	2	42	5	16	92	2	7	5	9	8	4	29	5	23	88	2	1	0	7	19	3	24	16	16
Benton Harbor	S.W.	90	25	9	2	3	7	6	17	12	9	83	23	4	7	2	8	6	17	5	11	74	6	4	2	3	15	7	23	3	11
Niles	S.W.	93	0	1	0	0	1	0	19	51	18	93	0	2	1	8	2	1	21	47	11	90	0	0	5	2	1	0	27	32	23
Battle Creek....	S. C.	93	0	3	9	0	8	4	24	24	21	93	3	5	11	2	3	15	21	25	8	90	1	5	3	5	10	23	13	21	9
Coldwater.......	S. C.	91	0	3	14	9	4	4	37	6	14	87	0	3	18	3	0	3	35	8	17	90	0	7	11	4	7	17	22	0	22
Kalamazoo......	S. C.	93	3	4	5	1	10	17	28	13	12	93	1	5	7	0	10	16	27	9	18	90	1	3	3	4	12	13	29	7	18
Mendon..........	S. C.	93	0	5	9	5	9	17	20	23	5	90	0	5	19	4	6	17	22	8	9	88	0	6	6	4	7	25	10	14	16
Tecumseh	S. C.	93	18	11	8	2	7	2	20	16	9	93	23	6	6	5	1	7	14	17	14	90	14	7	0	10	3	11	11	14	20
Ypsilanti........	S. C.	91	0	6	13	2	1	12	28	19	10	88	5	3	9	4	6	6	18	25	12	86	5	4	7	3	9	10	13	12	23
Detroit	S. E.	93	0	3	16	7	4	6	36	17	4	93	1	14	9	6	2	5	37	8	11	90	1	8	5	10	3	12	23	13	15
Woodmere Cem.	S. E.	93	3	8	9	5	0	18	17	26	7	93	10	8	9	4	1	15	15	17	14	90	4	4	6	6	1	10	17	27	15
Washington	S. E.	93	39	3	10	2	3	9	6	15	6	93	37	10	12	2	3	5	7	10	7	90	36	5	1	5	2	12	4	10	15

* The names of observers, their places of observation, and the counties in which these places are situated, are stated in Exhibit 7, page 306.

† The full names of the divisions and the counties in each division are stated in Exhibit 1, page 227.

‡ This line is an average for only the 16 stations from which statements nearly complete were received for every month of the year; it does not include the lines for Menominee, Agricultural College, Lansing, Ionia, Benton Harbor, Coldwater, or Ypsilanti.

Graphic representations of statements for 16 lines in this table are given in Diagram XIII., page 356, which is explained on pages 307–8.

TABLE XIII.—CONTINUED.—*Directions of Wind, Months in 1879.—Observations at which the Wind was Blowing from Directions Named.*

STATIONS IN MICHIGAN.*	DIVISIONS OF THE STATE.†	OCTOBER. Total.	Calm.	N.	N. E.	E.	S. E.	S.	S. W.	W.	N. W.
Average for 16 Stations....‡	......	93	7	4	4	3	10	16	22	16	11
Marquette......	U. P.†	93	3	3	0	0	5	16	31	12	23
Menominee......	U. P.	93	3	1	11	10	26	28	7	1	6
Petoskey.......	N.	93	5	2	3	0	4	11	35	8	25
Alpena.........	N. E.	93	3	2	1	1	23	9	16	16	17
Grand Haven...	W.	93	2	14	1	4	2	25	23	11	11
Nirvana.........	W.	93	0	2	14	9	14	3	34	9	8
Port Huron.....	B.&E.	93	2	7	4	3	3	38	13	16	7
Thornville......	B.&E.	93	23	0	7	1	18	5	21	15	3
Agrl. College...	C.	93	34	2	3	1	4	8	20	6	15
Ionia...........	C.	93	0	5	15	4	4	10	29	24	2
Lansing.........	C.	93	0	1	5	2	16	16	26	19	8
Otisville........	C.	92	8	3	3	2	18	11	19	23	5
Benton Harbor.	S. W.	......	......	..	....	..	....	..	.....	..	....
Niles............	S. W.	93	0	0	1	1	13	15	21	28	14
Battle Creek...	S. C.	93	0	3	5	7	7	24	11	34	2
Coldwater......	S. C.	93	1	0	2	8	16	23	16	9	18
Kalamazoo......	S. C.	93	0	2	4	0	17	21	30	8	11
Mendon.........	S. C.	87	0	7	4	4	8	32	12	5	15
Tecumseh......	S. C.	93	19	3	2	4	6	11	20	22	6
Ypsilanti.......	S. C.	88	7	0	10	1	12	8	25	14	11
Detroit..........	S. E.	93	3	3	4	9	6	9	29	23	7
Woodmere Cem.	S. E.	93	5	4	6	3	5	17	21	17	15
Washington....	S. E.	93	36	3	2	1	8	5	13	14	11

STATIONS IN MICHIGAN.*	DIVISIONS OF THE STATE.†	NOVEMBER. Total.	Calm.	N.	N. E.	E.	S. E.	S.	S. W.	W.	N. W.
Average for 16 Stations....‡	......	90	4	4	5	3	9	15	19	17	15
Marquette......	U. P.†	90	3	5	3	2	7	4	15	26	25
Menominee......	U. P.	80	1	10	21	10	17	9	3	4	5
Petoskey.......	N.	90	0	10	7	0	15	17	19	6	17
Alpena.........	N. E.	90	0	7	2	3	10	16	16	21	15
Grand Haven...	W.	90	0	13	6	18	8	12	11	15	7
Nirvana.........	W.	90	3	1	22	5	11	1	28	3	16
Port Huron.....	B.&E.	90	1	1	5	1	7	26	13	19	17
Thornville......	B.&E.	90	13	0	9	2	9	16	17	13	11
Agrl. College...	C.	......	......	..	......	..	....	..	....	..	....
Ionia...........	C.	90	0	5	10	0	4	16	28	18	9
Lansing.........	C.	90	0	1	3	3	14	10	26	19	14
Otisville........	C.	90	1	0	4	2	20	10	20	14	19
Benton Harbor.	S. W.	45	3	3	4	4	1	11	1	13	5
Niles............	S. W.	90	0	9	0	0	11	4	35	13	18
Battle Creek...	S. C.	90	0	1	4	7	6	24	10	31	7
Coldwater......	S. C.	90	0	0	2	2	11	7	36	10	22
Kalamazoo......	S. C.	90	0	1	3	1	11	20	34	5	15
Mendon.........	S. C.	90	0	2	3	2	5	33	15	21	9
Tecumseh......	S. C.	90	10	5	1	0	12	10	15	20	17
Ypsilanti.......	S. C.	88	6	0	7	2	9	6	36	15	7
Detroit..........	S. E.	90	2	7	0	3	4	11	28	21	14
Woodmere Cem.	S. E.	90	0	5	3	1	2	23	14	27	15
Washington....	S. E.	90	29	3	3	2	3	10	16	10	14

STATIONS IN MICHIGAN.*	DIVISIONS OF THE STATE.†	DECEMBER. Total.	Calm.	N.	N. E.	E.	S. E.	S.	S. W.	W.	N. W.
Average for 16 Stations....‡	......	92	4	4	8	7	11	8	20	18	13
Marquette......	U. P.†	93	2	2	6	2	10	8	10	26	27
Menominee......	U. P.	......	......	..	....	..	....	..	....	....	......
Petoskey.......	N.	93	1	1	6	2	6	7	26	9	35
Alpena.........	N. E.	93	0	10	6	4	11	5	15	23	19
Grand Haven...	W.	93	1	5	13	16	9	9	10	20	10
Nirvana.........	W.	93	4	1	14	10	21	0	27	6	10
Port Huron.....	B.&E.	93	2	5	7	12	9	11	22	18	7
Thornville......	B.&E.	93	10	4	8	4	18	5	19	18	7
Agrl. College...	C.	......	......	..	....	..	....	..	....	....	......
Ionia...........	C.	93	0	0	14	0	9	2	42	21	5
Lansing.........	C.	93	0	1	12	7	18	4	22	21	8
Otisville........	C.	91	2	4	3	7	19	2	19	17	18
Benton Harbor.	S. W.	92	12	5	5	6	11	16	9	24	7
Niles............	S. W.	93	0	5	0	0	6	5	30	30	17
Battle Creek...	S. C.	87	0	3	6	9	10	7	13	30	9
Coldwater......	S. C.	93	0	1	8	9	15	6	37	10	7
Kalamazoo......	S. C.	93	0	4	16	4	11	15	34	2	7
Mendon.........	S. C.	91	0	1	9	10	14	18	24	10	5
Tecumseh......	S. C.	93	8	2	7	8	14	6	16	21	11
Ypsilanti.......	S. C.	82	0	3	18	9	9	2	5	30	6
Detroit..........	S. E.	93	1	9	9	12	5	5	24	20	8
Woodmere Cem.	S. E.	93	1	8	8	11	1	14	12	29	9
Washington....	S. E.	93	28	7	7	6	4	4	19	10	8

* The names of observers, their places of observation, and the counties in which these places are situated, are stated in Exhibit 7, page 306.
† The full names of the divisions and the counties in each division are stated in Exhibit 1, page 227.
‡ This line is an average for only the 16 stations from which statements nearly complete were received for every month of the year; it does not include the lines for Menominee, Agricultural College, Lansing, Ionia, Benton Harbor, Coldwater, or Ypsilanti.

Graphic representations of statements for 16 lines in this table are given in Diagram XIII., page 356, which is explained on pages 307–8.

EXHIBIT 21.—*Comparison of the Average Atmospheric Pressure during the Year and during each Month of the Year 1879, with Averages for the 15 Years 1864–78; also with Averages for the four Years 1875–1878; and for the Year 1878,—Inches of Mercury in the Barometer, Corrected for Temperature and for Instrumental Error.—Observations made at 7 A. M., 2 P. M., and 9 P. M., Daily, by Prof. R. C. Kedzie at the State Agricultural College*, near Lansing, Michigan.*

YEARS, ETC.	AVERAGE ATMOSPHERIC PRESSURE,—INCHES OF MERCURY.												
	Annual Av.	Jan.	Feb.	Mar.	Apr.	May.	June.	July.	Aug.	Sept.	Oct.	Nov.*	Dec.*
Av. 15 Years—'64-78†	29.010	29.026	29.003	28.976	28.959	28.967	28.995	29.016	29.035	29.078	29.044	29.019	29.034
Av. 4 Years—'75-78†	29.016	29.047	29.010	28.933	28.962	29.023	28.986	29.038	29.020	29.064	29.014	29.035	29.059
1878	29.041	29.087	29.014	29.015	28.863	29.001	29.018	29.073	28.998	29.145	29.068	29.069	29.135
1879*	29.083	29.118	29.095	29.090	29.038	29.104	29.061	29.045	29.088	29.140	29.148	29.029	29.034
In 1879 **Higher** than Av. 15 Years, 1864–78	.073	.092	.092	.114	.079	.137	.066	.029	.053	.062	.104	.010	0
In 1879 **Higher** than Av. 4 Years, 1875–78	.067	.071	.085	.157	.076	.081	.075	.007	.068	.076	.134	-----	-----
In 1879 **Lower** than Av. 4 Years, 1875–78	--------	-----	-----	-----	-----	-----	-----	-----	-----	-----	-----	.006	.025
In 1879 **Higher** than in 1878	.042	.031	.081	.075	.175	.103	.043	-----	.090	-----	.080	-----	-----
In 1879 **Lower** than in 1878	--------	-----	-----	-----	-----	-----	-----	.028	-----	.005	-----	.040	.101

* For November and December, 1879, the observations were made by Harry B. Turner, at the office of the State Board of Health, Lansing.

† Pike's barometer was used prior to 1875, Green's standard barometer has been used since 1874. To the observations from 1864-1874 inclusive, .12 in. has been added for instrumental error. From those from 1875 to 1879 .012 in. has been subtracted for instrumental error.

The record of the barometer at the Agricultural College is higher than that for the State Board of Health in every month of the year.

October 25, highest barometer since I have kept record, 30.704 in. (corrected). [Probably corrected for temperature and reduced to sea level. The 7 A. M. observation simply corrected for temperature is 30.030 in.]—*Supt. F. W. Higgins, on Oct. register.*

TABLE XIV.—*Average Atmospheric Pressure, for the Year, and for each Month in the Year 1879, at 19 Stations in Michigan, as indicated by the Height, in inches, of Mercury in the Barometer. Corrected for Temperature,—Reduced to 32° F., (for some Stations not corrected for Instrumental Errors §).—Average¶ of Observations made Daily at 7 A. M., 2 P. M., and 9 P. M., by Observers* for the State Board of Health and for the U. S. Signal Service.*

STATIONS IN MICHIGAN.*	GEOGRAPHICAL DIVISIONS OF THE STATE.†	INCHES OF MERCURY,—ATMOSPHERIC PRESSURE.													
		YEAR.		MONTHS, 1879.											
		1878.	1879.	Jan.	Feb.	March.	April.	May.	June.	July.	August.	Sept.	Oct.	Nov.	Dec.
Average for 13 Stations ‡	-------	a	29.155	29.169	29.173	29.163	29.122	29.176	29.121	29.091	29.106	29.188	29.221	29.152	29.176
Marquette *, §	U. P. †	-------	29.267	29.249	29.323	29.272	29.310	29.311	29.233	29.189	29.215	29.275	29.271	29.260	29.299
Menominee	U. P.	-------	b	29.251	29.305	29.265	29.270	29.285	h 29.229	29.182	29.208	29.279	29.285	i 29.229	-------
Alpena §, ¶	N. E.	-------	29.340	29.312	29.369	29.362	29.329	29.379	29.317	29.277	29.307	29.377	29.397	29.267	29.387
Grand Haven §	W.	-------	29.359	29.385	29.383	29.300	29.337	29.373	29.331	29.301	29.316	29.403	29.439	29.364	29.372
Nirvana	W.	-------	29.054	29.049	29.070	29.080	29.036	29.069	29.026	28.987	29.013	29.095	29.120	29.043	29.057
Port Huron §	B. & E.	-------	29.341	29.334	29.341	29.359	29.296	29.365	29.303	29.276	29.290	29.384	29.423	29.349	29.372
Thornville	B. & E.	-------	c	-------	28.937	28.945	28.903	28.982	28.926	28.902	28.916	29.000	29.029	28.942	28.949
Agrl. College, near Lansing §	C.	29.053	d	29.118	29.095	29.090	29.038	29.104	29.061	29 045	29.088	29.140	29.148	-------	-------
Lansing §	C.	-------	29.045	j 29.057	29.045	29.015	k 29.004	l 29.085	29.016	28.998	29.030	29.106	29.115	29.029	29.034
Otisville	C.	29.040	e	28.703	28.692	28.698	28.649	28.700	-------	m 29.065	m 29.095	h 29.158	o 29.211	o 29.156	o 29.162
Benton Harbor	S. W.	29.482	f	o 29.596	j 29.609	l 29.602	j 29.540	-------	-------	k 29.535	l 29.539	n 29.608	-------	p 29.663	m 29.614
Battle Creek	S. C.	28.651	28.655	28.659	28.685	28.662	28.611	28.665	28.600	28.578	28.610	28 701	28.733	28 675	o 28.680
Kalamazoo	S. C.	29.037	29.099	29.121	29.103	29.112	29.059	29.112	o 29.057	29.043	29.055	29.136	29.175	29.103	29.106
Mendon	S. C.	29.011	g	k 29.695	k 29.299	k 29.256	l 29.120	o 29.040	o 28.791	m 28.694	o 28.774	h 28.897	k 28.886	o 28.880	o 28.860
Tecumseh	S. C.	-------	29.041	29.136	29.111	29.104	29.005	29.049	29.063	28.996	28.938	29.015	29.071	28.992	29.016
Ypsilanti	S. C.	-------	28.942	m 29.000	28.939	28.937	28.874	28.956	h 28.893	o 28.866	l 28.890	h 28.958	o 29.009	k 28.980	j 29.004
Detroit §, ¶	S. E.	29.268	29.348	29.362	29.355	29.362	29.292	29.366	29.297	29.281	29.295	29.395	29.433	29.365	29.375
Woodmere Cem., near Detroit §	S. E.	29.320	29.387	29.399	29.399	29.400	29.333	29.399	29.343	29.318	29.331	29.427	29.470	29.406	29.419
Washington	S. E.	-------	29.135	29.130	k 29.132	29.152	29.096	29.160	29.092	29.067	29.088	29.174	29.216	29.142	29.167

* The names of observers, their places of observation, and the counties in which these places are situated, are stated in Exhibit 7, page 306.

† The full names of the divisions and the counties in each division are stated in Exhibit 1, page 227.

‡ This line is an average for only 13 of the localities from which reports were received for every month of the year. It does not include the line for Mendon, as the barometer used at Mendon was not a standard barometer.

§ For stations marked thus (§) a correction has been made for instrumental error, as follows: For Marquette, .004 in. added; for Alpena, .043 subtracted; for Grand Haven, .002 added; for Port Huron, .001 added; for the Agricultural College, .012 subtracted; for Lansing, .002 added; for Detroit, .009 added till April 6, 1879, and .017 added for the rest of the year; for Woodmere Cemetery, .010 subtracted. For other stations the instrumental error of barometer is not known.

¶ The daily "means" from which the numbers in the lines for Alpena, Port Huron, and Detroit were computed at those stations, were found by multiplying the 9 P. M. observation by 2, adding the other two observations, and dividing the sum by 4. For the other stations, the daily averages were computed by dividing the sum of the three daily observations by 3.

a The average for 8 stations in 1878 is 29.108 in. b For 11 months, 29.253. c For 11 months, 28.948. d Including the observations at Lansing for Nov. and Dec., the average of this line is 29.083 in. e For 11 months, 28.935. f For 9 months, 29.589. g For 12 months, 29.016; the barometer used at Mendon in 1879 not being a standard barometer, the line for Mendon has not been included in the average line. The facts that the barometer at Mendon was moved into the house June 1, and that it has been reduced by the same tables as the other barometers, may explain its being so low from June to December, as compared with the first half of the year. h For 28 days. i For 24 days. j For 25 days. k For 27 days. l For 26 days. m For 30 days. n For 19 days. o For 29 days. p For last 15 days.

Graphic representations of 13 representative lines in this table are given in Diagram XIV., opposite this page.

DIAGRAM No. XIV.—ATMOSPHERIC PRESSURE, BY MONTHS, IN 1879.

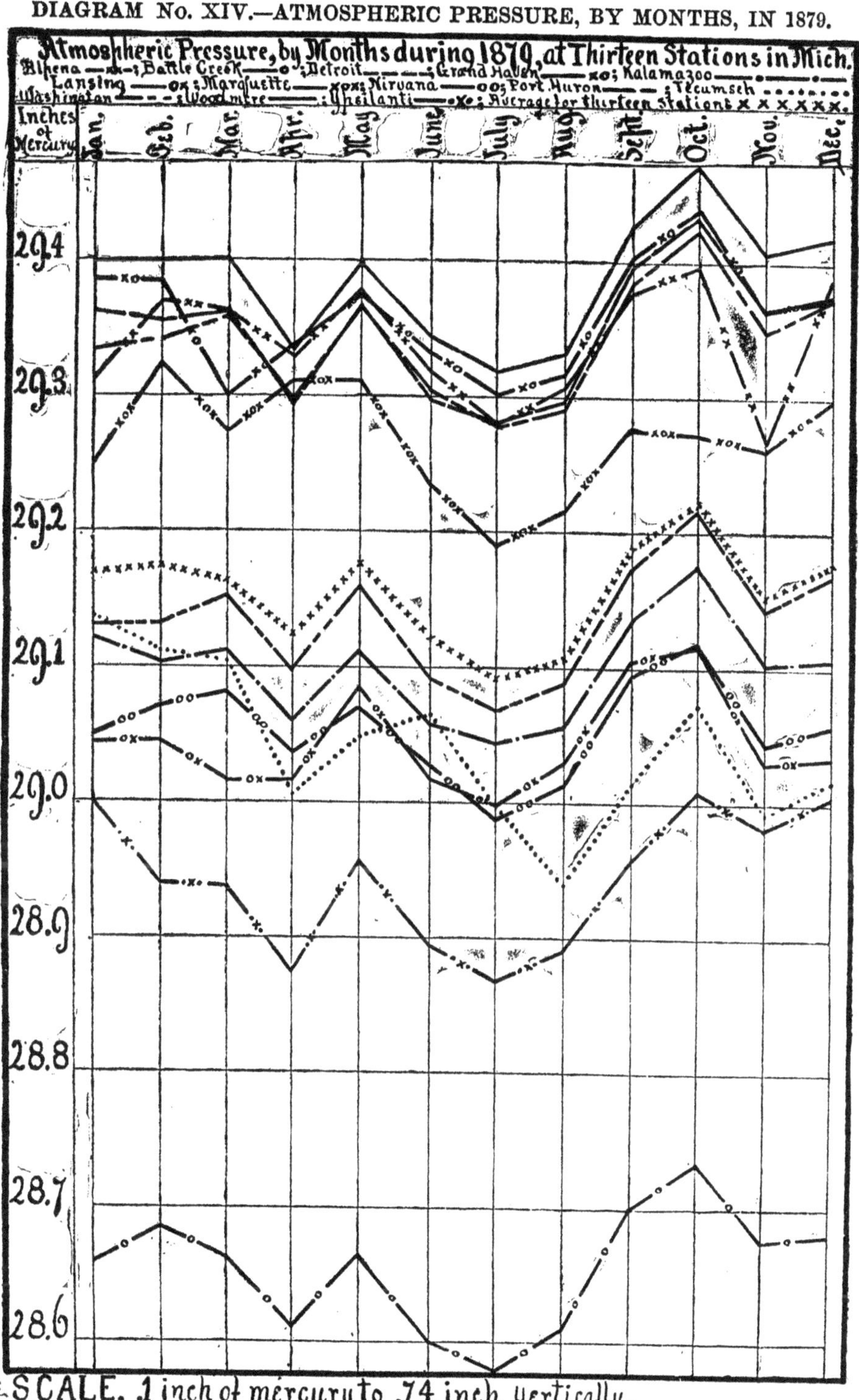

*SCALE, .1 inch of mercury to .74 inch vertically.

DESIGNED BY HENRY B. BAKER. DRAWN BY H. B. TURNER AND JNO. K. ALLEN.

www.ingramcontent.com/pod-product-compliance
Lightning Source LLC
LaVergne TN
LVHW010623110826
845149LV00003B/1028

9781418186845